COMBINED-CYCLE GAS & STEAM TURBINE POWER PLANTS

Second Edition

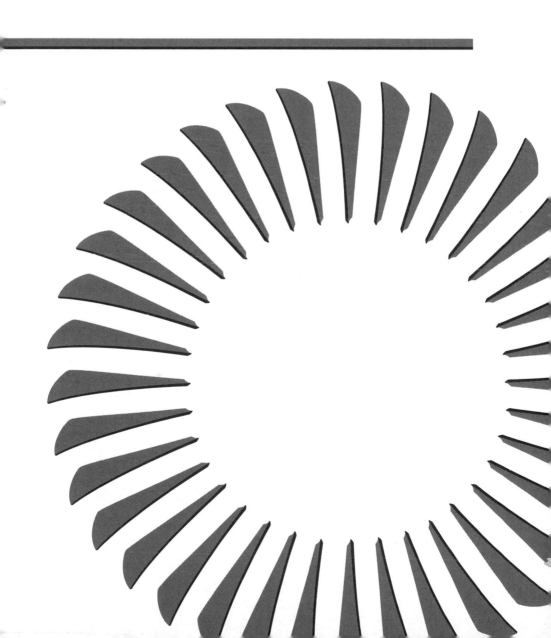

COMBINED-CYCLE GAS & STEAM TURBINE POWER PLANTS

Second Edition

by

**Rolf Kehlhofer, Rolf Bachmann,
Henrik Nielsen, Judy Warner**

Tulsa, Oklahoma

Copyright 1999 by
PennWell Publishing Company
1421 S. Sheridan Road/P.O. Box 1260
Tulsa, Oklahoma 74101

Library of Congress Cataloging-in-Publication Data
Kehlhofer, Rolf; Bachmann, Rolf; Nielsen, Henrik; Warner, Judy
      Combined Cycle Gas & Steam Turbine Power Plants/Kehlhofer, Bachmann,
Nielsen, Warner
          p.   cm.
      Includes index.
      ISBN 0-87814-736-5
      1.   2.
      I. Title

Printed in the United States of America.
    3 4 5 02 01

Contents

List of Figures and Tables

Tables

INTRODUCTION

CHAPTER 1

1

Introduction

When two thermal cycles are combined in a single power plant the efficiency that can be achieved is higher than that of one cycle alone. Thermal cycles with the same or with different working media can be combined; however, a combination of cycles with different working media is more interesting because their advantages can complement one another. Normally, when two cycles are combined, the cycle operating at the higher temperature level is called the "topping cycle". The waste heat it produces is then used in a second process that operates at a lower temperature level and is therefore called the "bottoming cycle".

Careful selection of the working media means that an overall process can be created which makes optimum thermodynamic use of the heat in the upper range of temperatures and returns waste heat to the environment at as low a temperature level as possible. Normally the topping and bottoming cycles are coupled in a heat exchanger.

The combination most widely accepted for commercial power generation is that of a gas topping cycle with a water/steam bottoming cycle. Figure 1-1 is a simplified flow diagram for such a cycle, in which the exhaust heat of a simple cycle gas turbine is used to generate steam for a steam process.

Replacement of the water/steam in this type of cycle with organic fluids or ammonia has been suggested in the literature because of potential advantages over water in the low exhaust-gas temperature range. However, as gas turbine exhaust temperatures are increased in line with gas turbine development, these advantages become insignificant compared to the high development costs and the potential hazard to the environment through problems such as ammonia leakage. These cycles do not appear likely ever to replace the steam process in a combined-cycle power plant.

The subject of this book is mainly the combination of an open-cycle gas turbine with a water/steam cycle. This combination—known commonly as the combined-cycle—has several advantages:

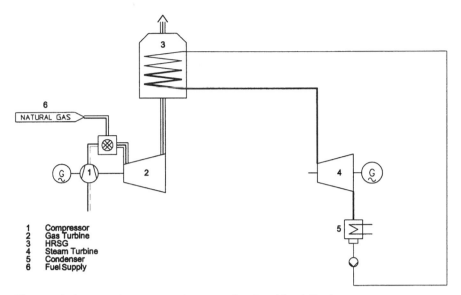

Figure 1-1 Simplified Flow Diagram of a Combined Cycle

- Air is a straight forward medium that can be used in modern gas turbines at high turbine inlet temperature levels (above 1,100°C (2,012°F)), providing the optimum prerequisites for a good topping cycle
- Steam/water is inexpensive, widely available, non-hazardous and suitable for medium and low temperature ranges, being ideal for the bottoming cycle

The initial breakthrough of these cycles onto the commercial power generation market was possible due to the development of the gas turbine. Only since the late 1970s have gas turbine inlet temperatures—and hence, exhaust-gas temperatures—been sufficiently high to design high-efficiency combined-cycles. This breakthrough was made easier because the components of the plant were not new, having already been proven in power plant applications with simple cycle gas turbines and steam turbine processes. This helped to keep development costs low. The result was a power plant with high efficiency, low installation cost and a fast delivery time.

THE ELECTRICITY MARKET

CHAPTER 2

2

The Electricity Market

Basic Requirements

The fundamental difference between electricity and most other commodities is that electricity can not be stored in a practical manner on a large scale. Storing electricity directly is very expensive and can only be done for small quantities (e.g., car batteries). Indirect storage through water or compressed air is more suitable for large-scale applications but it is very much dependent on topography and is in most cases not economical.

For this reason, electricity must be produced when the customers need it. It has to be transported by means of extensive transmission and distribution systems, which help to stabilize and equalize the load in the system. Nevertheless, large fluctuations in demand during the day require quick reactions from generation plants in order to maintain the balance between demand and production.

Fulfillment of this task has been the main focus of the industry from the beginning. A reliable supply of electricity, efficiently delivered, was and remains the major priority.

In the last few years a new priority has been set by a global trend to deregulate the electric power market. Deregulation means an opening to competition of what has been a generally closed and protected industry. Private investors have begun to install their own power plants and supply power to the grid. This has created a major focus shift: generators now have to compete in order to sell their product.

In the past–in a regulated environment–they sold the power they generated on a cost-plus basis. Production costs were of lower priority than the requirement of grid stability and reliability. They focused on reliability through extensive specifications and sufficient system redundancy. Costs could usually be transferred to the end-use customers. The new competitive situation has altered key success factors for the electricity generators.

Today, overall production cost *is* a key to their success. They must offer electricity at the lowest cost, yet still meet the requirement of flexible adjustment between demand and supply. This cost factor is of major importance for the so-called *merchant plants*. These are plants built by investors who accept the full market risk, expecting that their assets will be competitive cost-wise and that they will get a good return on their investment during the lifetime of the plant. Risk and its mitigation carry a much higher weight in this environment.

On one side of this equation, higher competitive risks must be taken in order to survive in the new markets. On the other side, risks such as cost and schedule overruns have to be minimized. These factors lead us to new behaviors, one of which is that electricity generators are buying plants for fixed, lump-sum prices with short throughput times. Constructing new plants–with long lead times and high capital costs–are more and more considered to be too risky from the investor's point of view.

Combined-cycle power plants benefit from this change. They have low investment costs and short construction times compared to large coal fired stations and even more so compared to nuclear plants.

The other benefits of combined-cycles are high efficiency and low environmental impact. Worldwide, levels of emissions of all kinds must meet stringent regulations acceptable to the public. It is therefore important for power producers to invest in plants with an inherently low level of emissions. Risk mitigation and public acceptance are paramount. Clean plants are easier to permit, to build and to operate. Combined-cycle plants–especially those fired with natural gas–are a good choice with their low emissions (see chapter 9 for more information about this point).

Throughout this book, we will review different combined-cycle applications, including cogeneration applications; although, more than 80% of the market concerns plants for pure power generation. We will examine how such plants affect and are affected by all of the variables discussed above.

The Global Power Generation Market

Some statistics about the power generation markets are presented in Tables 2-1 to 2-3 and Figure 2-1.

Table 2-1 World Installed Capacity by Technology and Fuel

Steam Turbine Power Plant	(66%)		
Fueled by:			
Coal		950	GW
Gas		330	GW
Oil		315	GW
Other		55	GW
Nuclear		360	GW
Steam Turbines for Combined-Cycle		40	GW
Total		**2050**	**GW**
Gas Turbine Power Plant	(10%)		
Gas Turbines for Simple Cycle		215	GW
Gas Turbines for Combined-Cycle		85	GW
Total		**300**	**GW**
Hydro Power Plant	(22%)		
Hydro		600	GW
Pumped storage		80	GW
Total		**680**	**GW**
Diesel Generator Power Plant	(2%)	**70**	**GW**
Total	**(100%)**	**3100**	**GW**

The data cited in Table 2-1 are valid for 1996. All figures are rounded to GW. Power generated by alternative fuels and renewable energy sources are excluded. Ratings considered for steam turbines, gas turbines and combined-cycles are larger than 3 MW. Ratings considered for diesels are larger than 1 MW. Approximately 40% of all gas turbines are powered by non-gas fuel. Table 2-2 shows a geographical breakdown of the world's power generation market by technology.

Table 2-2 World Power Generation by Technology

Electricity Generation by Source 1995	World excluding FSU & EE (TWh)	World excluding FSU & EE (%)	FSU & EE (TWh)	World with FSU & EE (TWh)
Hydro	2,247.9	20.2%		
Nuclear	2,051.4	18.5%		
Other	190.3	1.7%		
Thermal Plants	6,617.4	59.6%		
TOTAL	**11,107.0**	**100.0%**	**1,733.7**	**12,840.7**

Source: DRI/McGraw-Hill: *The Future of the Electric Power Industry World Overview*, February 1996, Standard & Poor's

Notes: FSU   =   Former Soviet Union
          EE   =   Eastern Europe
          TWh   =   1,000 GWh

Table 2-3 World Fuel Consumption for Fossil Fueled Power Plants

Fuel Inputs to Fossil Fueled Plants by Fuel for 1995	World excluding FSU & EE (TTOE)	World excluding FSU & EE (%)
Oil	233,180	14.6%
Gas	291,113	18.3%
Solid Fuels	1,070,602	67.1%
TOTAL	**1,594,895**	**100.0%**

Source: DRI/McGraw-Hill: *The Future of the Electric Power Industry World Overview*, February 1996, Standard & Poor's

Notes: FSU   =   Former Soviet Union
        EE   =   Eastern Europe
        TTOE   =   Thousand Tons of Oil Equivalent
        Solid Fuels =   Hard Coal, Coke, Brown Coal

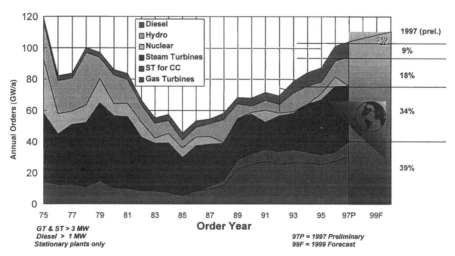

Figure 2-1 Market Development. Source: ABB

Data breakdown for FSU and EE are not available; however, the total electricity generation for 1995 for these regions is known.

In 1997, installed power capacity of combined-cycle plants worldwide totaled more than 150 GW with new installations averaging about 25 to 30 GW per year.

Cost of Electricity

The cost of electricity is a specific term related to MWh of electricity produced. It consists mainly of capital cost, fuel cost, and operation and maintenance costs. In the current trend towards deregulation of the power generation industry the cost of the generated electricity is a key element when selecting the type of power plant for a given application. Other factors that are evaluated include:

- permitting procedure
- financiability, loan structures
- environmental concerns (nuclear waste, air emissions, water consumption, heat rejection, noise)
- construction time, depreciation period of the project, etc

Every power plant is designed to keep the production cost as low as possible. Legislation and environmental protection give boundary conditions to this goal.

In the following pages, the cost of electricity for different types of power stations is determined. It should be noted that the presented information can vary depending on local and regional conditions, and therefore represents general trends (e.g., fuel gas prices can be regionally below half of world market prices making plants using fuel gas even more competitive).

Capital costs per unit of electricity for a given power plant depend on the price and the amortization rate for that plant, on interest or on the desired yield on capital investments (annuity factor), and on the load factor of the plant. Capital costs are also influenced by the interest during construction.

Fuel costs per unit of electricity are proportional to the specific price of the fuel and inversely proportional to the average electrical efficiency of the installation. This average electrical efficiency must not be mixed up with the electrical efficiency at rated load. It is defined as follows:

$$\overline{\eta} = \eta \cdot \eta_{Oper} \tag{2-1}$$

where:

η is the electrical efficiency at rated load. (This is the % of the fuel that is converted into electricity at rated load)

η_{Oper} is the operating efficiency, which takes into account the following losses:

- start-up and shutdown losses
- higher fuel consumption for part load operation
- other miscellaneous energy and heat losses (e.g., due to fouling, aging, operator error, etc.)

Operation and maintenance costs consist of fixed costs of operation, maintenance and administration (staff, insurance, etc.) and the variable costs of operation and maintenance, and repair (consumables, spare parts, etc.).

By adding the capital cost, fuel cost and operation and maintenance cost the **cost of electricity** is calculated. *Present value* is generally the basis used for economic comparisons. The various costs for a power station are incurred at different times but for financial calculations are corrected to a single reference time, which is generally the date on which commercial operation starts. These converted amounts are referred to as present value.

The cost of electricity (US$/MWh) is calculated as follows:

$$Y_{EL} = \frac{TCR \cdot \psi}{P \cdot T_{eq}} + \frac{Y_F}{\overline{\eta}} + \frac{U_{fix}}{P \cdot T_{eq}} + u_{var} \tag{2-2}$$

where:

TCR: Total capital requirement to be written off (current value of all expenses during planning, procurement, construction and commissioning such as the price of the plant, construction interest, etc.) (US $)

ψ Annuity factor: $\psi = \dfrac{q-1}{1-q^{-n}}$ [1/a]

P Rated power output in MW

T_{eq} Equivalent utilization time at rated power output, in hours/annum (h/a)

Y_F Price of fuel (US\$/MWh thermal = 3.412x US\$/MBTU)

η Average plant efficiency

U_{fix} Fixed cost of operation, maintenance and administration (US\$/a)

u_{var} Variable cost of operation, maintenance and repair (US\$/MWh)

q 1+z

z discount rate (%/a)

n amortization period in years

The equivalent utilization time at rated output is the electrical energy generated by a plant in a period of time divided by the rated output. This definition enables corrections to be made for the effects of different operating modes (e.g., part-load operation) for the power plants under consideration in an electrical grid, so that they can be analyzed on a comparable basis. For fuel cost and operation and maintenance cost, no escalation rates have been applied to calculate the cost of electricity.

The cost of electricity can be calculated by using the equivalent utilization time and the fuel price as variables, but it must be understood that the calculated cost of electricity is an average figure. In a deregulated power generation market, power stations do not quote on an average cost of electricity basis but on the basis of demand and supply. Therefore, it is important to understand that the above equation contains fixed and variable costs.

Fixed costs are:

- interest and depreciation on capital
- the fixed costs of operation, maintenance and administration (e.g., staff)

Variable costs are:

- the fuel used
- the variable costs of operation, maintenance and repair (e.g., spare parts)

For a time of low demand and high supply (e.g., night hours), power stations can quote a price as low as the variable costs and, for short periods of time, an even lower figure, since stopping a station also incurs costs. During times of high demand (e.g., noon peak) they can quote at a level which will recuperate additional fixed costs. For reasons of simplicity, the average cost of electricity will be used for the following comparisons.

Competitive Standing of Combined-Cycle Power Plants

On the following pages, the combined-cycle power plant is compared with other thermal plants. The comparison evaluates the following types of power stations:

- diesel generator plants
- steam turbine plants
- gas turbine plants
- nuclear plants
- combined-cycle plants

The main range of ratings under consideration is between 30 and 1000 MW. Combined-cycles with a smaller output can, of course, be built but they are less interesting for pure power generation because the relative cost increases as the power rating decreases. These are optimally used for heat and power production (e.g., district heating or process-steam delivery at the same time as power generation).

Comparison of turnkey prices

Figure 2-2 shows how specific investment costs for the various types of power plants depend on the power output. These costs are valid for turnkey installations. They are based on 1998 price levels and progress payments and do not include interest during construction. The data shown indicates trends, so appropriate caution must be taken in applying them, since many factors affect the price of a power plant–site-related factors such as soil conditions; earthquake factors; noise limits; types of cooling and corresponding structures; emission limits; labor rates; commercial risks; legal regulations, and so forth. Figure 2-2 also shows the low investment costs required for the gas

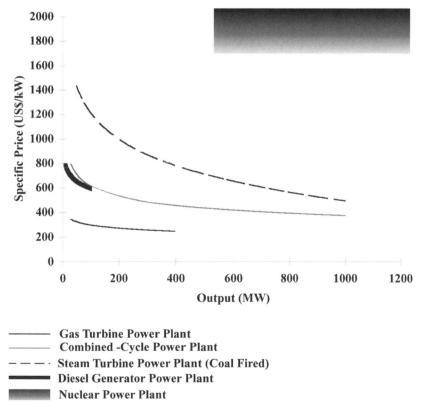

———— Gas Turbine Power Plant
———— Combined -Cycle Power Plant
– – – · Steam Turbine Power Plant (Coal Fired)
▬▬ Diesel Generator Power Plant
▬▬ Nuclear Power Plant

Figure 2-2 Comparison of Different Turnkey Power Plants in Terms of Specific Price and Output

turbine, which have contributed significantly to its wide spread acceptance.

Taken together with its simplicity and short start-up time to full load, the gas turbine is an attractive peak-load machine. Steam power plants are more expensive than combined-cycle power plants. A coal-fired plant, for example, costs approximately two times more than a combined-cycle plant with the same output.

Nuclear power plants are very expensive and have a wide range of investment costs due to the strict emphasis on safety standards, legislation requirements and other local considerations. Combined-cycle plants are quite inexpensive and therefore easier to finance compared to conventional steam and nuclear plants, which makes off-balance sheet financing possible for investors.

Figure 2-3 shows the breakdown of the total cost in a combined-cycle plant between the main equipment.

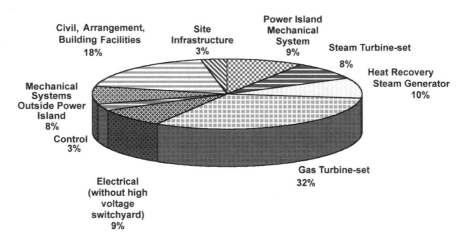

Figure 2-3 The Cost Percentage of the Different Plant Areas for a Typical 400 MW Turnkey Combined Cycle Plant

Comparison of efficiency and fuel costs

At today's fuel prices, efficiency is an important factor for installations operated at intermediate or base load. If an expensive fuel like liquefied natural gas (LNG) is used, the efficiency is crucial. For that reason high efficiency is a prerequisite for having an economical plant. The net efficiency of a power station is defined as:

$$\eta = \frac{P_{net}}{HI} \qquad (2\text{-}3)$$

where:

P_{net} Power output at the high voltage terminals of the step-up transformer in MW. This number considers the power consumption of all the plant auxiliaries

HI   Heat input to the power station in MJ/s measured at the plant boundary (i.e., mass flow of the fuel multiplied by the lower heating value of the fuel)

Fuel that contains hydrogen produces water vapor when burned. If the combustion products are cooled to the point where all of this vapor is condensed, then the maximum possible heat is extracted defining the higher heating value (HHV) of the fuel. In practice, this latent heat cannot be used and the heat extracted is the lower heating value (LHV). The considerations in the following chapters arc always based on the LHV of the fuel.

Figure 2-4 indicates how electrical efficiency at rated load for the different types of powcr plant relates to the power output. Steam turbine power plants have been segregated into coal fired and nuclear plants. The combined-cycle plants are without supplementary firing. The chart points out the thermodynamic superiority of the combined-cycle plant. This was made possible, to a large extent, by gas turbine technology which already achieves an efficiency of 38% to 40% with a turbine inlet temperature of 1300°C (2372°F). Only a few years ago, the efficiency of a newly installed coal-fired steam power plant was at these levels–but with much a higher investment cost and complexity! Coal-fired power plants can remain competitive where coal is cheap or natural gas is not available.

A nuclear power station has a lower efficiency due to the very low live steam conditions, which are saturated steam at approximately 60 bar (860 psig). This results in heat rejection to the environment that is approximately two and a half times more than that of a newly built combined-cycle plant with the same electrical output.

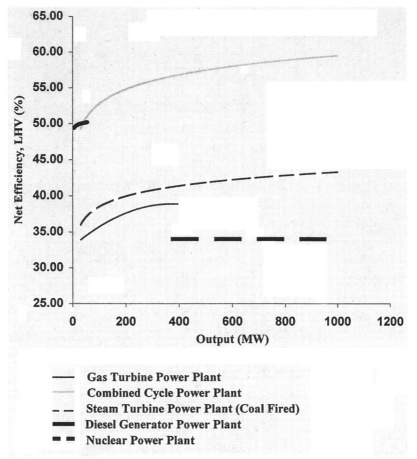

Figure 2-4 Net Efficiencies for GT, CC, ST (Coal Fired), Nuclear and Diesel Power Plants

When determining the specific fuel costs of power generation, efficiency is one factor; the other is the price of fuel–specifically, the fuel cost portion in the cost of electricity is the ratio of fuel price and efficiency. A power plant can remain competitive despite a low efficiency, when the fuel used is cheap.

Fuel selection and the corresponding type of power plant is determined not only by short-term economic considerations, but also in accordance with political criteria and assumptions about long-term developments in the prices of the various fuels available. In this regard, the following aspects are important in selecting the type of power station to be built:

- long-term availability of the fuel at a competitive price
- short-term alternative for the primary fuel
- risk of supply shortages due to political interference
- opposition (e.g., the political discussion about nuclear technology)
- environmental considerations that favor a clean fuel like natural gas
- independence from a single fuel source
- strategic reasons to use a domestic fuel
- financing requirements (e.g., uninterruptible fuel supply)

Table 2-4 lists fuels that are burned in thermal power plants today.

Table 2-4 Fuel Flexibility of Various Power Plants

Fuel	Gas Turbine	Combined-Cycle	Steam Power Plant	Diesel Generator
Natural Gas/LNG	Yes	Yes	Uneconomical	Yes
Diesel	Yes	Yes	Uneconomical	Yes
Crude Oil/Heavy Fuel	Yes$_1$	Yes$_1$	Yes	Yes
Coal	No	No	Yes	No
Refuse	No	No	Yes	No
Coal Gas, Industry Gas, Low Heat Content Gas	Yes$_2$	Yes$_2$	Yes	Yes
Nuclear Fuel	No	No	Yes	No

$_1$ Heavy Oil or Crude Oil can be burned if the gas turbine is designed accordingly

$_2$ These fuels can be used in gas turbines. Modifications to the gas turbine are necessary for fuels with a low heating value

LNG stands for liquefied natural gas

Some gas turbines can burn heavy oil or crude oil. Heavy-duty gas turbines designed for industry are more suitable for this than those derived from jet technology (aero derivatives). Gas turbines with large combustion chambers and single burners are better capable of burning heavy fuels than those with several burners/combustion chambers since the latter are more sensitive to changes in flame length, radiation, etc. An additional requirement for burning heavy oil or crude oil in a gas turbine is the correct treatment of the fuel, generally by means of cleaning the fuel and/or dosing it with additives. These steps make it possible to remove or inhibit elements that cause high temperature corrosion, such as vanadium and sodium.

Modern gas turbines with high firing temperatures are, in general, not designed for heavy fuel operation, but mainly for natural gas and distillate oil.

As can be seen from Figure 2-2 and Figure 2-5, combined-cycle plants are low in capital cost but burn an expensive fuel compared to coal fired steam turbine and nuclear plants. The biggest contribution in the cost of the electricity generated is the fuel cost. Long-term supply agreements or partnerships with fuel companies, or equity investments by fuel suppliers into the plant, are possible ways to cope with the risk of wide swings in the fuel prices. As the electricity market becomes deregulated, the fuel market is opening. When spot market gas prices are high, owners of a modern combined-cycle plant with an uninterruptible fuel-gas contract can switch to the back-up fuel and sell the saved gas on the spot market. In this way they may take advantage of a high demand situation and attractive back-up fuel prices, thus generating additional profit for the owner. Fuel flexibility is higher for a steam turbine power plant than a combined-cycle power plant. But combined-cycle plants are superior to steam power plants for power generation when gas or diesel is used, due to higher levels of efficiency, lower investments, and more attractive operating costs.

Fossil fuel prices are volatile (with the exception of coal) and therefore need due consideration for a long-term investment. While it is possible to include the estimated fuel-price escalation during the scheduled service life of the unit in the calculations of economic costs, such an estimate should be treated with caution.

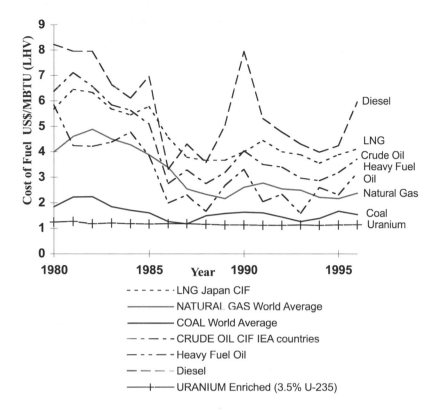

Figure 2-5 The Cost of Fuels. Source: Cost Insurance Freight BP Statistical Review of World Energy.

Uranium, the fuel for nuclear power stations, and coal, the main fuel for steam power plants, are quite stable at low costs. Combined-cycle plants mainly use natural gas that is more expensive (though it has become cheaper over the years) but convert it into electricity with a high efficiency. This makes the combined-cycle plant more competitive for power generation and increases its share of the industry. Current known world resources for both coal and gas are expected to last for more than 200 years and 60 years respectively at the average world wide consumption rate.

Comparison of operation and maintenance costs

At current levels of fuel and capital cost, operation and maintenance costs affect the economy of a power plant in a limited manner only. They strongly depend on site specific and local conditions and account for approximately a tenth of the cost of electricity in a combined cycle plant. Figure 2-6 shows the variable operation and maintenance cost of the different power plants. Variable costs for a combined-cycle plant are lower than for gas turbine plants because these costs are driven by the spare parts of the gas turbine, which can be distributed over a larger output in the combined-cycle plant.

Figure 2-7 shows the fixed operation and maintenance costs of the different power plants.

Nuclear power stations and conventional steam turbine power plants require staff over and above what is required for other kinds of

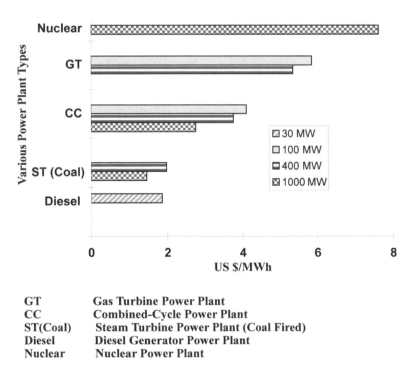

GT	Gas Turbine Power Plant
CC	Combined-Cycle Power Plant
ST(Coal)	Steam Turbine Power Plant (Coal Fired)
Diesel	Diesel Generator Power Plant
Nuclear	Nuclear Power Plant

Figure 2-6 Variable Operating and Maintenance Costs for Various Power Plants of Different Sizes. Fuel costs not included

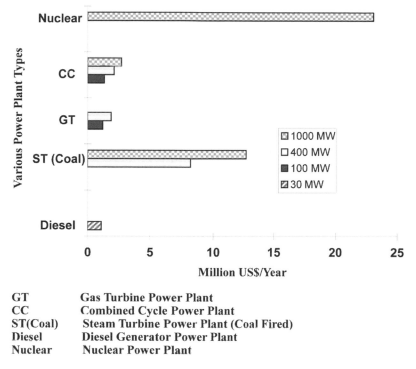

GT	Gas Turbine Power Plant
CC	Combined Cycle Power Plant
ST(Coal)	Steam Turbine Power Plant (Coal Fired)
Diesel	Diesel Generator Power Plant
Nuclear	Nuclear Power Plant

Figure 2-7 Fixed Operating and Maintenance Costs for Various Power Plants of Different Sizes

plants, which is another cost driver for fixed operation and maintenance costs.

Comparison of availability and reliability

The availability of a power plant is defined as:

$$AF = \frac{PH - SOH - FOH}{PH} \qquad (2\text{-}4)$$

where:

PH hours of the period, normally one year, which amounts to 8,760 h

SOH scheduled outage hours for planned maintenance

FOH forced outage hours for unplanned outages and repairs

The reliability of a power plant is defined as:

$$RF = \frac{PH - FOH}{PH} \tag{2-5}$$

So reliability is the percentage of the time between planned overhauls where the plant is ready to answer the call, whereas the availability is the percentage of total time where power could be produced.

Availability and reliability have a big impact on plant economy. When a unit is down, power must either be generated in another power station or purchased from another producer. In each case, replacement power is more expensive. The power station's fixed costs are incurred whether the plant is running or not.

In deregulated markets, reliability is crucial. At peak tariff hours, a major portion of the income is generated and the fixed costs can be written off. Scheduled outages can be planned for off-peak periods when tariffs are close to or even below variable costs. Then only a small loss of income results from the planned outages.

In weak electrical grids or grids with few interconnections–and for plants with power purchase agreements that contain capacity payments–availability is also important.

No values for reliability can be stated that will be valid for all cases since factors such as the fuel used, preventive maintenance, and operating mode have an impact. However, statistics indicate that all types of plants under consideration have similar availabilities and reliabilities when operated under the same conditions. Typical figures for the availability and reliability of well designed and maintained plants are detailed in Table 2-5.

Table 2-5 Availability and Reliability of Generating Plants

Type of Plant	Availability	Reliability
Gas Turbine Plant (gas fired)	88-95%	97-99%
Steam Turbine Plant (coal fired)	82-89%	94-97%
Combined-Cycle Plant (gas fired)	86-93%	95-98%
Nuclear Power Plant	80-89%	92-98%
Diesel Generator (diesel fired)	90-95%	96-98%

These figures are valid for plants operated at base load. They would be lower for peak or intermediate-load plants because frequent start-ups and shutdowns reduce lifetime and increase the scheduled maintenance and forced outage rates.

Major factors determining plant availability are:

- design of the major components
- engineering of the plant as whole, especially of the interfaces between the systems
- mode of operation (whether base-, intermediate-, or peak-load duty)
- type of fuel
- qualifications and skill of the operating and maintenance staff
- adherence to manufacturer's operating and maintenance instructions

Availability is not used in the cost-of-electricity calculation because the equivalent utilization time is the variable. However, it must be considered that a high availability allows an operator to run a power plant with a higher utilization time per year and therefore achieve a higher income.

Comparison of construction time

The time required for construction affects the economics of a unit–the longer it takes, the larger the capital employed without return, since construction interest, insurance and taxes during the construction period add to the price of the plant.

Figure 2-8 shows the length of time required to build the various types of power plant.

A gas turbine in a simple-cycle application can be installed within the shortest time frame because of its standardized design. Gas turbines therefore help secure power generation in fast-growing economies. Additional time is needed for the completion of a combined-cycle plant. Combined-cycle plants can be installed in two phases–phased installation, with the gas turbine running first in sim-

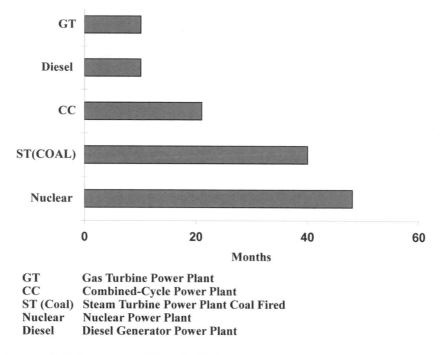

GT   Gas Turbine Power Plant
CC   Combined-Cycle Power Plant
ST (Coal) Steam Turbine Power Plant Coal Fired
Nuclear   Nuclear Power Plant
Diesel   Diesel Generator Power Plant

Figure 2-8 Construction Time for Various Power Plants (From 'Notice To Proceed' to 'Commercial Operation')

ple-cycle mode, and then in combined-cycle mode as the steam cycle becomes available. With this procedure, two-thirds of the power is available in the time required for a gas turbine installation. However, an outage is needed to convert the gas turbine power plant from simple-cycle to combined-cycle mode. This procedure was often applied in the past. Today, with the short installation time of a modern combined-cycle plant, phased installation is seldom attractive.

Cost of Electricity Comparison

Based on the data presented in Figures 2-2 through 2-8, and the equations in this chapter, the cost of electricity has been calculated. Figures 2-9 to 2-14 show the effects of the most important parameters on the cost of electricity of a power plant. Utilization times used in Fig-

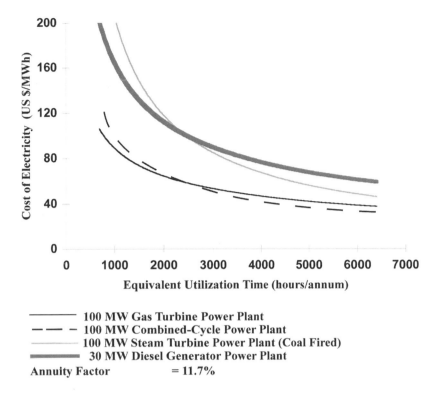

Figure 2-9 Dependence of the Cost of Electricity on the Equivalent Utilization Time

ures 2-9 through 2-11 were corrected with the reliability of the individual types of plant to reflect that a forced outage disturbs operation and causes start-up and shut down losses and additional wear and tear on the equipment.

The following conclusions can be drawn from these diagrams:

The main competition to the combined-cycle comes from gas turbine and coal-fired steam turbine power plants. This situation is unlikely to change in the near future.

For small- to medium-power outputs (up to approximately 30 MW), a diesel generator power plant can be a genuine alternative. The high efficiency of modern diesel engines is slightly less than combined-cycle with the same rating. To achieve a higher output with diesel generators, however, multiple units must be combined. There-

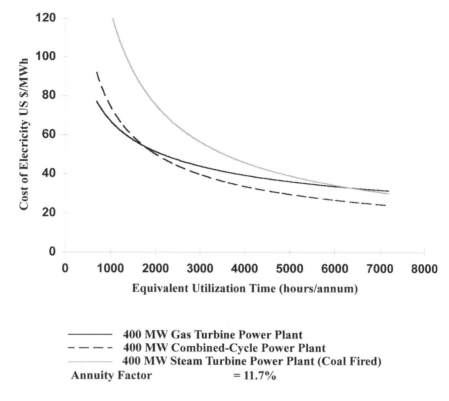

Figure 2-10 Dependence of the Cost of Electricity on the Equivalent Utilization Time

fore, the diesel-based plant loses its attractiveness for higher power ratings, because investment costs are higher than those for combined-cycle power plants without compensating for that fact by providing greater fuel flexibility. The diesel engine is also less desirable for environmental reasons. It is more difficult to attain low emission levels with it, particularly for NOx and unburned hydrocarbons.

Conventional steam power plants are suitable for use as coal burning plants operating in base-load (or occasional intermediate-load duty) if cheap coal is available or gas is expensive (e.g., LNG) for a combined cycle plant. Whenever gas or oil is fired in a power station, the combined-cycle plant is more economical than the steam power plant due to its higher efficiency and lower specific price. Modern combined-cycle plants are simpler, less expensive, and operationally more flexible than steam power plants.

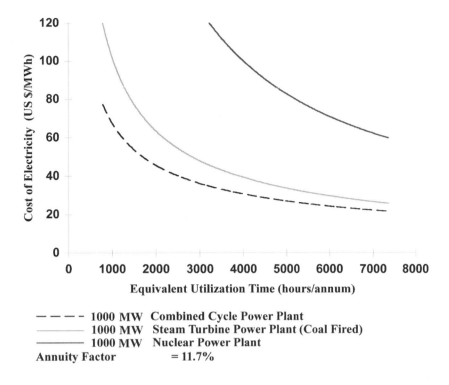

Figure 2-11 Dependence of the Cost of Electricity on the Equivalent Utilization Time

The choice between a steam power plant and combined-cycle plant for intermediate-to-base-load applications is a question of fuel. If natural gas is available, a combined-cycle design will be selected. If coal is the fuel, a steam power plant will be chosen.

Nuclear power is the most expensive power. The cost of the power generated makes nuclear power plants unsuitable for a fully deregulated power generation market. Due to very high capital investments and long construction times, depreciation periods of several decades result with the corresponding risk to the lender. Nuclear power can be competitive where gas prices are very high (e.g., countries which depend on LNG) and coal is not available at low cost (e.g., expensive transport).

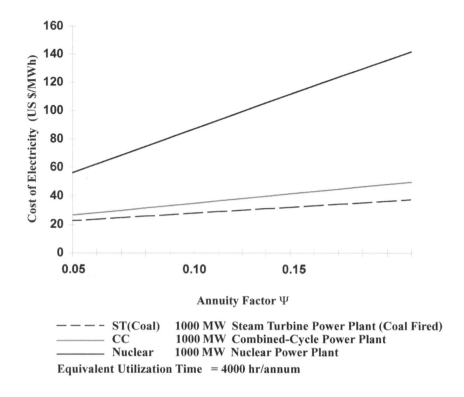

---- ST(Coal)   1000 MW **Steam Turbine Power Plant (Coal Fired)**
——— CC   1000 MW **Combined-Cycle Power Plant**
——— Nuclear   1000 MW **Nuclear Power Plant**
Equivalent Utilization Time = 4000 hr/annum

Figure 2-12 Dependence of Cost of Electricity on the Annuity Factor

For short utilization periods (peaking units), the gas turbine is most economical. Gas turbines can serve as intermediate- or base-load units in countries where fuel is abundant at low cost. The lack of water consumption has made this machine popular in dry regions. The short installation time allows a customer to plan a new installation on short notice.

If all fuels are readily available at world market prices, gas fired combined-cycle plants are the most economical solution for intermediate- and base-load applications. This results in a limited environmental impact (small heat rejection or low water consumption). With clean fuels like natural gas, this technology also achieves low emissions. Additional arguments in favor of a combined-cycle power plant are the ease of permitting for a clean, gas-fired plant, as opposed to a coal-burning unit; the shorter construction time with lower risk; and the

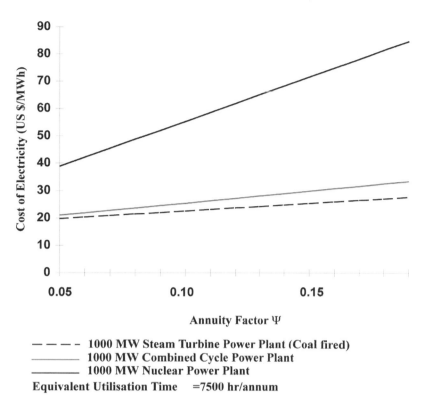

Figure 2-13 Dependence of Cost of Electricity on the Annuity Factor

smaller amount of CO_2 released per MWh electricity generated, which is well below half that of a modern coal fired unit.

Combined-cycle plants will remain competitive due to the large resources of natural gas and the increasing pipeline grid.

Efficiency improvements

As has been noted, combined-cycle power plants are often fired with natural gas which is more expensive than uranium or coal. But the specific price for a combined-cycle power plant is relatively low and the efficiency relatively high. The question answered below is how much can additionally be invested in a combined-cycle plant to gain additional electrical efficiency?

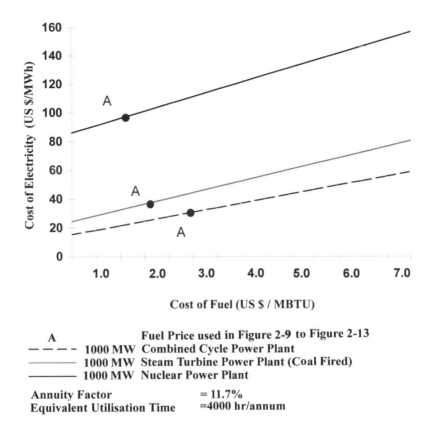

Figure legend:

A   **Fuel Price used in Figure 2-9 to Figure 2-13**
– – – –   **1000 MW Combined Cycle Power Plant**
─────   **1000 MW Steam Turbine Power Plant (Coal Fired)**
─────   **1000 MW Nuclear Power Plant**

Annuity Factor   **= 11.7%**
Equivalent Utilisation Time   **=4000 hr/annum**

Figure 2-14 Dependence of Cost of Electricity on the Cost of Fuel

Some efficiency improvements can be found in the water/steam cycle, which has a direct impact on the steam turbine output. This means that a 1% increase in electrical efficiency of the plant results in a 1% output increase because the fuel input to the gas turbine (and therefore, to the plant) remains constant. The maximum additional capital that can be invested to gain a 1% increase in efficiency is given by the limit that the cost of electricity remains constant.

Using the equation to determine the cost of electricity, the following calculations can be done:

Index 1   is used for the plant that is used as the basis
Index 2   is used for the plant with a 1% improved efficiency

$$Y_{EL} = \frac{TCR_1 \cdot \psi}{P_1 \cdot T_{eq}} + \frac{Y_F}{\eta_1} + \frac{U_{FIX1}}{P_1 \cdot T_{eq}} + u_{var1} =$$

$$\frac{TCR_2 \cdot \psi}{1.01 \cdot P_1 \cdot T_{eq}} + \frac{Y_F}{1.01 \cdot \eta_1} + \frac{U_{FIX2}}{1.01 \cdot P_1 \cdot T_{eq}} + u_{var2} \qquad (2\text{-}6)$$

Because operation and maintenance costs are only approximately a tenth of the cost of electricity and remain constant for both plants, the above equation can be simplified to:

$$\frac{TCR_1 \cdot \psi}{P_1 \cdot T_{eq}} + \frac{Y_F}{\eta_1} \approx \frac{TCR_2 \cdot \psi}{1.01 \cdot P_1 \cdot T_{eq}} + \frac{Y_F}{1.01 \cdot \eta_1} \qquad (2\text{-}7)$$

which can be solved for TCR_2.

$$TCR_2 \approx TCR_1 \cdot 1.01 + \left(\frac{Y_F}{\eta_1} - \frac{Y_F}{1.01 \cdot \eta_1} \right) \frac{1.01 \cdot P_1 \cdot T_{eq}}{\psi} \qquad (2\text{-}8)$$

with:

TCR_1	=	400 million US $
Y_F	=	8.3 US $ / MWh (thermal) or 2.4 US $ / MBTU
η_1	=	55%
ψ	=	11.7 %
P_1	=	1000 MW
T_{eq}	=	7000 h/a
TCR_2	$\approx$	413.1 million US $ = 1.033 x TCR_1

For a 1% efficiency increase, 3.3% more capital can be invested for this example. It can be seen from above equation that this percentage will increase if the fuel price is higher or a lower discount rate is used. A smaller percentage results for fewer operating hours per year or a shorter depreciation period.

These considerations are essential in defining a combined-cycle plant for a given application.

THERMODYNAMIC PRINCIPLES OF THE COMBINED-CYCLE PLANT

CHAPTER **3**

3

Thermodynamic Principles of the Combined-Cycle Plant

Basic Considerations

The Carnot efficiency is the efficiency of an ideal thermal process:

$$\eta_c = \frac{T_E - T_A}{T_E} \tag{3-1}$$

where:

η_c = Carnot efficiency
T_E = Temperature of the Energy supplied, [K]
T_A = Ambient Temperature, [K]

Naturally, the efficiencies of real processes are lower since there are losses involved. A distinction is drawn between energetic and exergetic losses. *Energetic losses* are mainly heat losses, and are thus energy that is lost from the process. *Exergetic losses* are internal losses caused by irreversible processes in accordance with the second law of thermodynamics.

The process efficiency can be improved by raising the maximum temperature in the cycle, releasing the waste heat at a lower temperature, or by improving the process to minimize the internal exergetic losses.

The interest in combined-cycles arises particularly from these considerations. By its nature, no single-cycle can make all of these improvements to an equal extent. It thus seems reasonable to combine two cycles–one with high process temperatures and the other with a good "cold end".

In a simple-cycle gas turbine, attainable process temperatures are high as energy is supplied directly to the cycle without heat exchange. The exhaust heat temperature, however, is also high. In the steam cycle, the maximum process temperature is much lower than the gas turbine process, but the exhaust heat is returned to the environment at a low temperature. As illustrated in Table 3-1, combining a gas turbine and a steam turbine thus offers the best possible basis for a high efficiency thermal process.

The last line in the table shows the Carnot efficiencies of the various processes (i.e., the efficiencies that would be attainable if the processes took place without internal exergetic losses). Although that is not the case in reality, this figure can be used as an indicator of the quality of a thermal process. The value shown makes clear just how interesting the combined-cycle power plant is when compared to processes with only one cycle. Even a sophisticated, supercritical conventional reheat steam turbine power plant has a Carnot efficiency around 20 points lower than that of a good combined-cycle plant.

For combined-cycle power plants, actual efficiencies are around 75% of the Carnot efficiency, whereas for conventional steam power plants, this figure is around 80%. The differences between the actual efficiencies attained by a combined-cycle power plant and the other processes are therefore not quite as large as illustrated in Table 3-1. The relatively larger drop in the combined-cycle efficiency is caused by higher internal energy losses duc to the temperature differential for exchanging heat between the gas turbine exhaust and the water/steam cycle.

Table 3-1 Thermodynamic Comparison of Gas Turbine, Steam Turbine and Combined-Cycle Processes

	GT	ST	CC
Average temperature of	1,000 – 1,350	640 - 700	1,000 – 1,350
heat supplied, K (°R)	(1,800 – 2,430)	(1,152 – 1,260)	(1,800 – 2,430)
Average temperature of	550 – 600	300 - 350	300 – 350
dissipated heat, K (°R)	(900 – 1,080)	(540 – 630)	(540 – 630)
Carnot efficiency, %	45 – 50	45 - 57	65 – 78

GT = Gas Turbine Power Plant,
ST = Steam Turbine Power Plant,
CC = Combined-Cycle Power Plant

As shown in Figure 3-1–which compares the temperature/entropy diagrams of the processes–the combined-cycle best utilizes the temperature differential in the heat supplied despite an additional exergetic loss between the gas and the steam process.

Efficiency of the Combined-Cycle Plant

It has been assumed until now that fuel energy is being supplied to the cycle only in the gas turbine. There are also combined-cycle in-

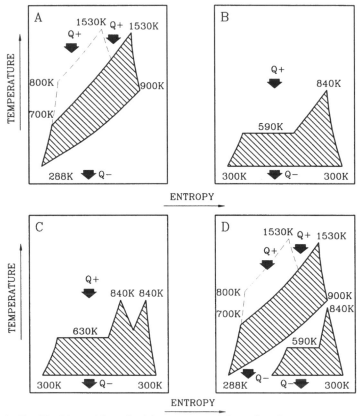

A **Gas Turbine with and without sequential combustion**
B **Steam Turbine without reheat**
C **Steam Turbine with reheat**
D **Combined cycle showing Gas Turbine topping and**
   Steam Turbine bottoming Cycle

Figure 3-1 Temperature/Entropy Diagrams for Various Cycles

stallations with additional firing in the heat recovery steam generator (HRSG), in which a portion of the heat is supplied directly to the steam process.

Accordingly, the general definition of the electrical efficiency of a combined-cycle plant is:

$$\eta_{cc} = \frac{P_{GT} + P_{ST}}{Q_{GT} + Q_{SF}} \qquad (3\text{-}2)$$

where:

P_{GT} = Gas turbine output

P_{ST} = Steam turbine output

Q_{GT} = Gas turbine fuel input

Q_{SF} = Additional / supplementary firing fuel input

This equation shows the so called gross efficiency of the combined-cycle because no station service power consumption and electrical losses, also called auxiliary consumption, P_{Aux}, have been deducted. If station auxiliary consumption is considered the net efficiency of the combined-cycle is given by:

$$\eta_{cc,net} = \frac{P_{GT} + P_{ST} - P_{Aux}}{Q_{GT} + Q_{SF}} \qquad (3\text{-}3)$$

In general, the efficiencies of the simple cycle gas and steam turbine processes can be defined in a similar manner:

$$\eta_{GT} = \frac{P_{GT}}{Q_{GT}} \qquad (3\text{-}4)$$

$$\eta_{ST} = \frac{P_{ST}}{Q_{GT,Exh} + Q_{SF}} \qquad (3\text{-}5)$$

where:

$$Q_{GT.Exh} \cong Q_{GT}(1 - \eta_{GT}) \qquad (3\text{-}6)$$

Combining these two equations yields:

$$\eta_{ST} = \frac{P_{ST}}{Q_{GT}(1 - \eta_{GT}) + Q_{SF}} \qquad (3\text{-}7)$$

This equation expresses the steam process efficiency of the combined-cycle.

If there is no supplementary firing in the HRSG, then equations 3-2 through 3-7 can be simplified by eliminating Q_{SF}, ($Q_{SF} = 0$). In view of the earlier considerations, it is generally better to burn the fuel directly in a modern gas turbine rather than in the HRSG because the temperature level at which heat is supplied to the process is higher (GT versus ST process in Table 3-1). For that reason interest in supplementary firing is decreasing.

The factors involved in combined-cycle installations with supplementary firing are discussed in more detail in chapter 4.

Efficiency of Combined-Cycles without Supplementary Firing in the HRSG

The most common and straightforward type of combined-cycle is one in which fuel is supplied in the gas turbine combustion chamber without additional heat supplied in the HRSG. By substituting equations 3-4 and 3-7 into equation 3-2:

$$\eta_{cc} = \frac{\eta_{GT} \cdot Q_{GT} + \eta_{ST} \cdot Q_{GT}(1 - \eta_{GT})}{Q_{GT}}$$

$$= \eta_{GT} + \eta_{ST}(1 - \eta_{GT}) \qquad (3\text{-}8)$$

Differentiation makes it possible to estimate the effect that a change in efficiency of the gas turbine has on overall efficiency:

$$\frac{\partial \eta_{cc}}{\partial \eta_{GT}} = 1 + \frac{\partial \eta_{ST}}{\partial \eta_{GT}} (1 - \eta_{GT}) - \eta_{ST} \qquad (3\text{-}9)$$

Increasing the gas turbine efficiency improves the overall efficiency only if:

$$\frac{\partial \eta_{cc}}{\partial \eta_{GT}} > 0 \qquad (3\text{-}10)$$

From Equation (3-9):

$$-\frac{\partial \eta_{ST}}{\partial \eta_{GT}} < \frac{1 - \eta_{ST}}{1 - \eta_{GT}} \qquad (3\text{-}11)$$

Improving the gas turbine efficiency is helpful only if it does not cause too great a drop in the efficiency of the steam process.

Table 3-2 indicates that when the efficiency of the gas turbine is raised, the reduction in efficiency of the steam process may be greater. (i.e., the gas turbine efficiency must be raised from 30 to 35% if the steam turbine efficiency is reduced from 30 to 27.8% (30/1.08) in order to keep the same overall combined cycle efficiency. However, under some circumstances both efficiencies can be raised at the same time). A gas turbine with maximum efficiency does not always provide an optimum combined-cycle plant. For example: For a gas turbine with single-stage combustion at constant turbine inlet temperature, a very-high pressure ratio attains a higher efficiency than a gas turbine with a moderate-pressure ratio. However, the efficiency of the combined-cycle plant with the second machine is normally better because the steam turbine operates far more efficiently with the higher exhaust gas temperature.

Table 3-2 Allowable Reduction in Steam Process Efficiency as a Function of Gas Turbine Efficiency ($\eta_{ST} = 0.3$)

η_{GT}	0.3	0.35	0.4
$-\dfrac{\partial \eta_{ST}}{\partial \eta_{GT}}$	1.0	1.08	1.17

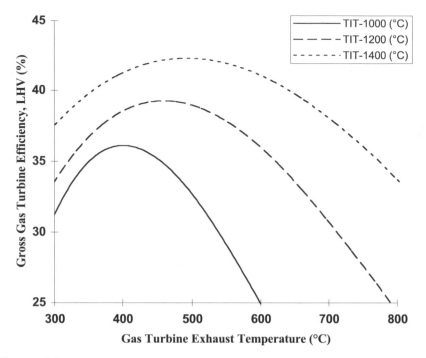

Figure 3-2a The Efficiency of a Simple Cycle Gas Turbine with Single Stage Combustion as a Function of Turbine Inlet Temperature (TIT) and the Turbine Exhaust Temperature

Figure 3-2a demonstrates the efficiency of the simple-cycle gas turbine with single stage combustion as a function of the turbine inlet and exhaust gas temperatures. The maximum efficiency is reached when the exhaust gas temperatures are quite low. In this case a low exhaust temperature is equivalent to a high-pressure ratio.

Figure 3-2b shows the overall efficiency of the combined-cycle based on the same gas turbine. Compared to Figure 3-2a, the optimum point has shifted toward higher exhaust temperatures from the gas turbine, which again indicate an over-proportional improvement of the water/steam cycle compared to the loss in gas turbine efficiency. For economic reasons, current gas turbines are generally optimized with respect to maximum power density (output per unit air flow) rather than efficiency. Often, this optimum coincides fairly accurately with the optimum efficiency of the combined-cycle plant. As a result, most of today's gas turbines are optimally suited for combined-cycle installations.

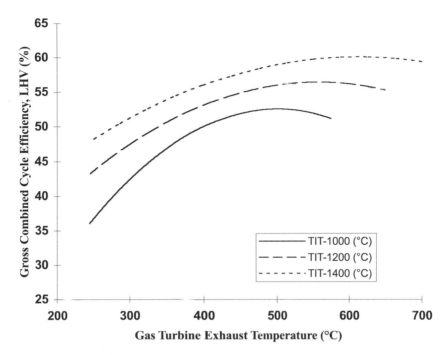

Figure 3-2b The Efficiency of a Combined-Cycle with a Single Stage Combustion Gas Turbine as a Function of the Turbine Inlet Temperature (TIT) and the Turbine Exhaust Temperature

Gas turbines of a more complicated design (i.e., with intermediate cooling in the compressor or recuperator) are less suitable for applications in combined-cycle plant. They normally have a high simple-cycle efficiency combined with a low exhaust gas temperature, so that the efficiency of the water/steam cycle is accordingly lower. These machines are more suitable for simple-cycle operation.

The effects of a high-pressure ratio and low combined-cycle efficiency can be uncoupled if the gas turbine is designed with sequential combustion- (air, upon leaving the compressor, is passed through the first combustion chamber and expands in the first turbine stage before final combustion and expansion). Gas turbines with sequential combustion have practically the same simple-cycle efficiencies as single-combustion gas turbines at the same overall pressure ratio and turbine inlet temperature.

For comparison, the same curves shown in Figure 3-2a and 3-2b are shown for gas turbines with sequential combustion in Figure 3-3a

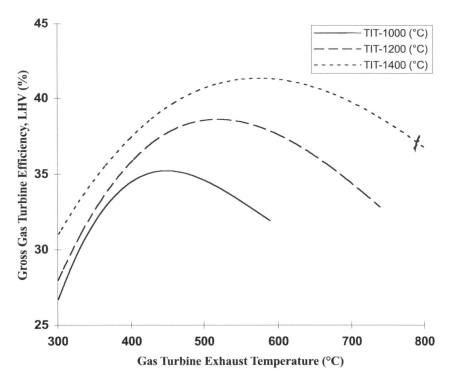

Figure 3-3a The Efficiency of a Simple Cycle Gas Turbine with Sequential Combustion as a Function of the Turbine Inlet Temperature (TIT) and the Turbine Exhaust Temperature

and 3-3b. Figure 3-3a shows almost the same optimum simple-cycle gas turbine efficiency level for the same gas turbine inlet temperatures (TIT). However, the exhaust gas temperatures are substantially higher, clearly improving the combined-cycle efficiency levels of Figure 3-3b compared to 3-2b. As for Figure 3-2a, a low exhaust gas temperature is equivalent to a high-pressure ratio of the sequential part (low-pressure part) of the gas turbine. For consistency, the pressure ratio of the high-pressure turbine is kept constant at 1.7:1.

The main advantage of a sequentially fired gas turbine is that the drawback of single- combustion gas turbines (i.e., pressure ratio and exhaust gas temperatures) is eliminated through the reheat process. This gives the ideal base for improved combined-cycle efficiencies, which also fits well to the Carnot comparison.

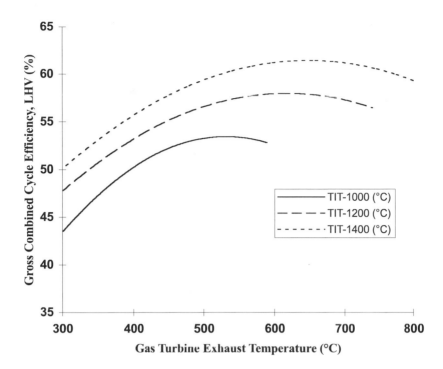

Figure 3-3b The Efficiency of a Combined-Cycle with a Sequential Combustion Gas Turbine as a Function of the Turbine Inlet Temperature (TIT) and the Turbine Exhaust Temperature

In summary, then, it may be said that the optimum gas turbine for simple-cycle and the optimum gas turbine for combined-cycle are not the same. The gas turbine with the highest efficiency does not necessarily produce the best overall efficiency of the combined-cycle plant. The type of gas turbine (i.e., gas turbine concept) and turbine inlet temperature are important factors. The gas turbine is generally a "standard machine" and must therefore be optimized by the manufactures for its main application (i.e., combined-cycle or simple-cycle).

More will be said about this in the following chapter.

COMBINED-CYCLE CONCEPTS

CHAPTER 4

4

Combined-Cycle Concepts

The main challenge in designing a combined-cycle plant with a given gas turbine is how to transfer gas turbine exhaust heat to the water/steam cycle to achieve optimum steam turbine output. The focus is on the heat recovery steam generator (HRSG) in which the heat transfer between the gas cycle and the water/steam cycle takes place.

Figure 4-1 shows the energy exchange that would take place in an idealized heat exchanger in which the product, mass flow times specific heat capacity, or the energy transferred per unit temperature must be the same in both media at any given point to prevent energy and exergy losses. In order for energy transfer to take place there must be a temperature difference between the two media. As this temperature difference tends towards zero the heat transfer surface of the heat exchanger tends towards infinity and the exergy losses towards zero. The heat transfer in an HRSG entails losses associated with three main factors:

- The physical properties of the water, steam and exhaust gases do not match causing exergetic and energetic losses
- The heat transfer surface cannot be infinitely large
- The temperature of the feedwater must be high enough to prevent corrosive acids forming in the exhaust gas where it comes into contact with the cold tubes. This limits the energy utilization by limiting the temperature to which the exhaust gas can be cooled

The extent to which these losses can be minimized (and the heat utilization maximized) depends on the concept and on the main parameters of the cycle. In a more complex cycle the heat will generally be used more efficiently, improving the performance but also increasing the cost. In practice, a compromise between performance and cost must always be made.

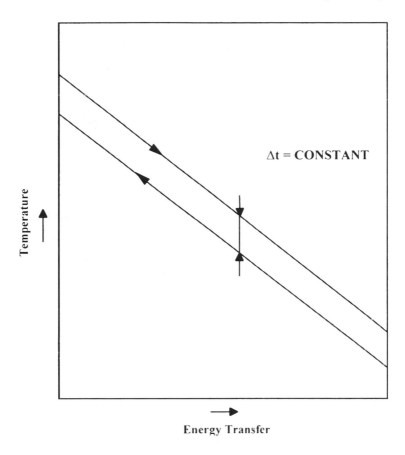

Figure 4-1 Energy/Temperature Diagram for an Idealized Heat Exchanger

Basic Combined-Cycle Concepts

In this section, the most common combined-cycle concepts are presented and explained, starting with the most simple and leading to more complex cycles. A heat balance for each of the main cycle concepts is given, based on ISO conditions (ambient temperature 15°C, (59°F); ambient pressure 1.013 bar, (14.7 psia); relative humidity 60%; condenser vacuum 0.045 bar (1.3"Hg)); with one ABB GT24 sequential combustion gas turbine, rated at 178 MW and a steam turbine with water cooled condenser. The gas turbine is equipped with cooling air coolers that generate additional steam for the water/steam cycle and boost the steam turbine output. Due to the fact that these features are the same for all of the heat balances, a clear comparison can be made between them showing how the cycle concept influences the heat utilization.

Additionally, an analysis is done for each concept to show how the main cycle parameters–such as live-steam temperatures and pressures and HRSG design parameters–influence the performance of the cycle. When designing a combined-cycle with a given gas turbine, the free parameters are those of the steam cycle. Therefore, the influence of the various parameters is analyzed with respect to the steam turbine output. This is because, in a plant without additional firing, the thermal energy supplied to the steam process is given by the gas turbine exhaust gas and the efficiency of the steam process is always proportional to the steam turbine output. The steam turbine does, however, account for only about 30% to 40% of the total combined-cycle power output so optimization of the steam process can only influence that portion.

Although the cycles shown have only one gas turbine, the concepts and results are also valid for cycles with several gas turbines and HRSGs of the same size.

Single-Pressure Cycle

The simplest type of combined-cycle is a basic single-pressure cycle, so called because the HRSG generates steam for the steam turbine at only one pressure level. A typical flow diagram (Fig. 4-2) shows a gas turbine exhausting into a single HRSG. The steam turbine (7) has a steam turbine bypass around it (8) into the condenser, which is used to accommodate the steam if for any reason it cannot be admitted to the steam turbine (e.g., during start-up or because the steam turbine is out of operation). After the condenser (9), a condensate pump (10) is used to pump the condensate back to the feedwater tank/deaerator (11). The feedwater pump (12) returns the feedwater to the HRSG. Heating steam for the deaerator is extracted from the steam turbine with a pegging steam supply (14) from the HRSG drum in case the pressure at the steam turbine bleed point becomes too low at off-design conditions.

The HRSG consists of three heat exchanger sections: the economizer (5), the evaporating loop (4), and the superheater (3). In the economizer, the feedwater is heated to a temperature close to its satu-

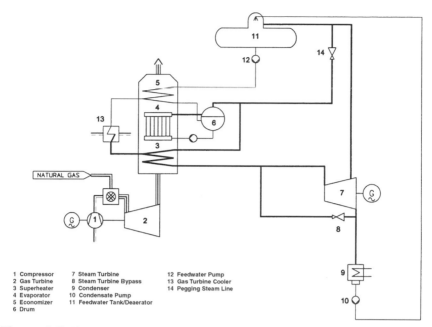

Figure 4-2 Flow Diagram of a Single Pressure Cycle

1 Compressor	7 Steam Turbine	12 Feedwater Pump
2 Gas Turbine	8 Steam Turbine Bypass	13 Gas Turbine Cooler
3 Superheater	9 Condenser	14 Pegging Steam Line
4 Evaporator	10 Condensate Pump	
5 Economizer	11 Feedwater Tank/Deaerator	
6 Drum		

ration point. The heated feedwater is evaporated at constant temperature and pressure in the evaporating loop. The water and saturated steam are separated in the drum (6) and the steam is fed to the superheater where it is superheated to the desired live-steam temperature.

Figure 4-3 shows the temperature/energy diagram for the single-pressure HRSG. The heat exchange in these three different sections is clearly recognizable. It is far removed from an idealized heat exchanger, mainly because of the fact that water evaporates at a constant temperature. The area between the gas and water/steam lines illustrates the exergy loss between the exhaust gas and the water/steam cycle. Even with an infinitely large heat transfer surface, this exergy loss can never be equal to zero and the heat exchange process in a boiler can never be ideal.

Two important parameters defining the HRSG are marked on the diagram. The approach temperature is the difference between the saturation temperature in the drum and the water temperature at the economizer outlet. This difference–typically 5 to 12 K (9 to 22°R)–helps to avoid evaporation in the economizer at off-design conditions.

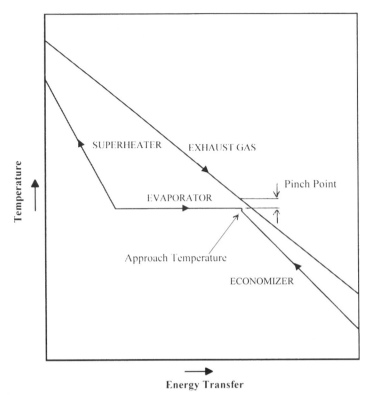

Figure 4-3 Energy/Temperature Diagram for a Single Pressure HRSG

The pinch-point temperature is the difference between the evaporator outlet temperature on the water/steam side and the on the exhaust gas side. This is important in defining the heating surface and performance of the HRSG. The lower the pinch-point the more heating surface is required and the more steam is generated. Pinch-points are typically between 8 and 15 K (14 to 27°R) depending on the economic parameters of the plant.

Figure 4-4 shows the heat balance for the single-pressure cycle where 73.3 kg/s (579 400 lb/hr) steam at 105 bar (1,508 psig) and 568°C (1,054°F) is generated. A loss of temperature and pressure in the live-steam line is seen, after which the steam is expanded in a steam turbine with an output of 94.8 MW. The resulting gross electrical efficiency of the cycle is 57.7%.

Energetic utilization of the exhaust heat is relatively low, considering that the feedwater temperature is 60°C (140°F), as illustrated

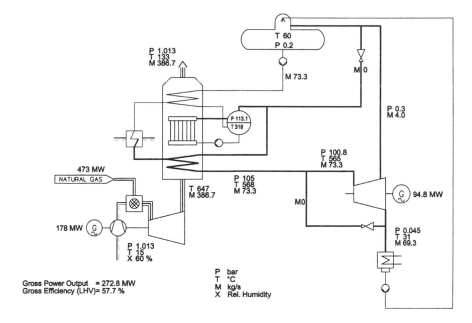

Figure 4-4 Heat Balance for a Single Pressure Cycle

by the stack temperature of 133°C (271°F). Fig. 4-5 shows this in an energy flow diagram, 11.4 % of the fuel energy supplied is lost through the stack. Another 29.9% is discharged in the condenser and 1% is equipment losses.

Main Design Parameters of the Single-Pressure Cycle

Live-steam pressure

In a combined-cycle plant, a high live-steam pressure does not necessarily mean a high efficiency of the cycle. Figure 4-6 shows how steam turbine output and HRSG efficiency vary with live-steam pressure. Expanding the steam at a higher live-steam pressure will give a higher steam turbine output due to the greater enthalpy drop in the steam turbine. However, due to the higher evaporation temperature less steam will be generated, resulting in a higher exhaust-gas temperature and a lower HRSG efficiency. An optimum is normally to be

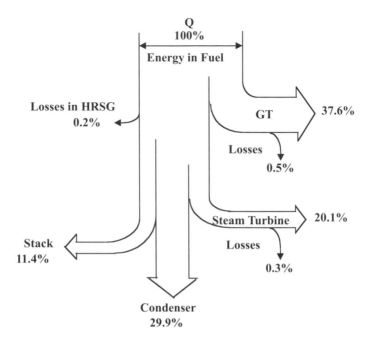

Figure 4-5 Energy Flow Diagram for the Single Pressure Combined-Cycle Plant

found between these two influences. In the example shown, the enthalpy gain in the steam turbine is dominant.

It is interesting that the high HRSG efficiency does not correspond to the high steam-turbine output. At a lower live-steam pressure there is a lower stack temperature and the energy in the exhaust gas is well-utilized. However, the steam turbine output at this live-steam pressure is lower because of the higher exergy losses in the HRSG. The reverse happens at higher live-steam pressures. This shows how exergy is more dominant than energy in determining the steam turbine output.

This effect is illustrated in Figure 4-7–the energy/temperature diagram for two different single-pressure cycles with live-steam pressures of 40 and 105 bar (566 and 1,508 psig). At the lower live-steam pressure, more thermal energy is available for evaporation and superheating, since the evaporation temperature is correspondingly lower. The pinch-point of the evaporator is the same in both cases. As a result, the stack temperature at 40 bar is about 11°C (20°F) lower than at 105 bar, which means that more waste heat energy is being utilized.

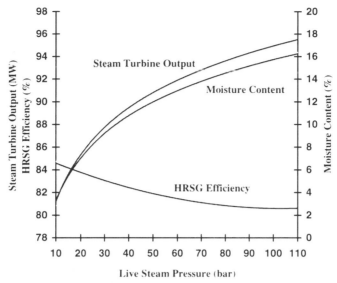

Figure 4-6 Effect of Live Steam Pressure on Steam Turbine Output, Steam Turbine Exhaust Moisture Content and HRSG Efficiency for a Single Pressure Cycle

One negative aspect of a higher live-steam pressure in a single-pressure cycle is an increase in the moisture content at the end of the steam turbine. Too much moisture increases the risk of erosion taking place in the last stages of the turbine. A limit is set at about 16%.

A change in the live-steam pressure affects the amount of heat to be removed in the condenser (as shown in Fig. 4-8) because of the change in the steam mass flow. The higher the live-steam mass flow the more heat must be removed in the condenser. The following–which arise due to an increase in the live-steam pressure–will bring economical advantages:

- a smaller exhaust section in the steam turbine
- smaller volume flows leading to smaller live-steam piping and valve dimensions
- a smaller condenser
- a reduction of the cooling water requirement

This can lead to considerably lower costs especially for power plants with expensive air cooled condensers.

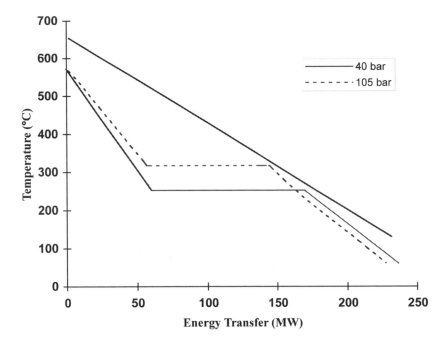

Figure 4-7 Energy/Temperature Diagram of a Single Pressure HRSG with Live Steam Pressures of 40 and 105 bar, (566 and 1,508 psig)

In general the total amount of live-steam can also influence the optimum live-steam pressure because it influences steam-turbine efficiency. Increasing the amount of live-steam will result in a larger volume flow and hence longer blades in the first row of the turbine, reducing the secondary blade losses. It follows that the optimum live-steam pressure also depends on the type of gas turbine used because different exhaust gas conditions will result in different live-steam flows for a given live-steam pressure. However, in the example, a pressure of 105 bar (1,508 psig) is chosen because it is the highest pressure possible without excessive moisture content in the steam turbine exhaust.

Live-steam temperature

For the chosen live-steam pressure, raising the live-steam temperature causes a very slight decrease in steam turbine output (as

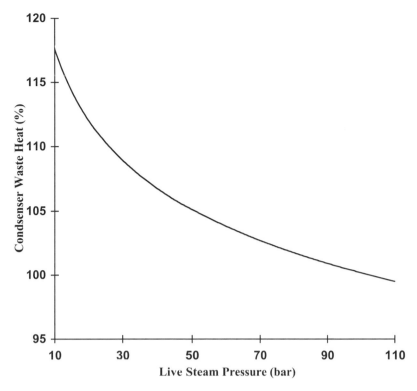

Figure 4-8 Effect of Live Steam Pressure on Condenser Waste Heat at Constant Vacuum

shown in Fig. 4-9). This is the result of two opposite effects. First, increasing the live-steam temperature, as when live-steam pressure is increased, results in a greater enthalpy drop in the steam turbine. Secondly, however, at the same time, additional superheating removes energy which would otherwise be used for steam production, resulting in a lower steam flow, a corresponding loss in the steam turbine output, and an increase in the stack temperature. The latter is the dominant effect in this case.

The live-steam temperature cannot be reduced below a certain limit for a given live-steam pressure because of the resulting increase in the moisture content in the steam turbine. In order to use a lower live-steam temperature, the pressure would have to be reduced, but this has a more negative effect on the steam turbine output than raising the live-steam temperature.

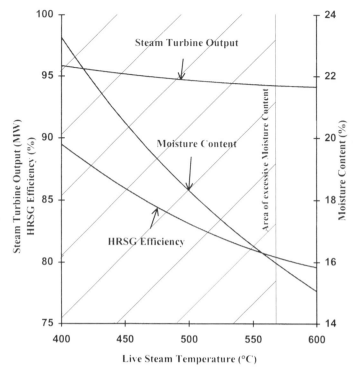

Figure 4-9 Effect of Live Steam Temperature on Steam Turbine Output, HRSG Efficiency and Steam Turbine Exhaust Moisture Content for a Single Pressure Cycle at 105 bar (1,508 psig) Live Steam Pressure

In the example, with 105 bar (1,508 psig) live-steam pressure, a temperature of 568°C (1,054°F) has been chosen which is well within the upper limit for current practice.

For gas turbines with lower exhaust-gas temperatures than are shown in the example, a lower live-steam pressure level would have to be chosen for the reasons given above.

Design Parameters of the HRSG

An important parameter in the optimization of a steam cycle is the pinch-point of the HRSG (defined in Figure 3-2), which directly affects the amount of steam generated. Figure 4-10 shows how, by reducing the pinch-point (with other parameters held constant), steam turbine output can be increased. This is due to a better rate of heat uti-

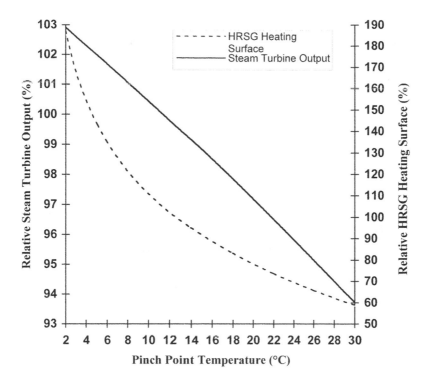

Figure 4-10 Effect of Pinch Point on Relative Steam Turbine Output and Relative HRSG Heating Surface

lization in the HRSG. However, the surface of the heat exchanger–and hence the cost of the HRSG–increases exponentially as the pinch-point tends towards zero, which sets a limit. In the example shown a pinch-point of 12 K (21.6°R) is used.

Similarly, a smaller approach-point temperature leads to a better heat utilization but increased HRSG surface. For a drum-type HRSG, a lower limit is set on the approach temperature by the need to minimize steaming in the economizers for off-design points. Our examples use 5 K (9°R) for the approach temperature.

It is the sum of the pinch-point and approach temperatures which determines the steam production in the HRSG for a given live-steam pressure and temperature. This means that a HRSG with a 10 K (18°R) pinch-point and 5 K (9°R) approach temperature would have the same steam production as one with 5 K (9°R) pinch-point and 10

K (18°R), approach temperature. However, the HRSG surface will not necessarily be the same in each case and so the optimum must be found, bearing in mind the steaming in the economizers and the impact of the HRSG surface.

A further influence on the energy available for evaporation and superheating is the steam-side pressure drop in the superheater. A higher pressure drop, for a given live-steam pressure, means that the evaporation takes place at a correspondingly higher pressure and temperature level where less energy is available for steam production. Other pressure losses, such as those in the economizers, do not influence the steam production but will have an effect on the power consumed by the feedwater pumps.

The design of the HRSG should be such that the pressure loss on its exhaust gas side, or back pressure, remains as low as possible. This loss strongly affects the power output and efficiency of the gas turbine because an increase in the back pressure will reduce the enthalpy drop in the gas turbine. Some of the lost output is, however, recovered in the water/steam cycle due to an increased gas turbine exhaust gas temperature. This effect is shown in Figure 4-11. The HRSG surface is higher for the lower pressure loss to account for the worse heat transfer caused by the lower exhaust gas velocity over the tube bundles. Typical HRSG pressure losses are 25 to 30 mbar (9.9 to 11.8 "W.C.). In the example 25 mbar is used.

Feedwater Preheating

In order to attain a good rate of exhaust gas heat utilization, the temperature of the feedwater should be kept as low as possible. Figure 4-12 demonstrates how, for cycles with only one stage of preheating using a steam turbine extraction output and efficiency fall sharply as the feedwater temperature is increased. This is because exhaust gases can, ideally, only be cooled to a temperature of between 10 K to 15 K (18 to 27°R) above the feedwater temperature. The higher the feedwater temperature, the hotter the gases being exhausted to the atmosphere through the stack and the more energy is wasted.

This is one significant difference between a conventional steam plant and the steam process in a combined-cycle plant with a high

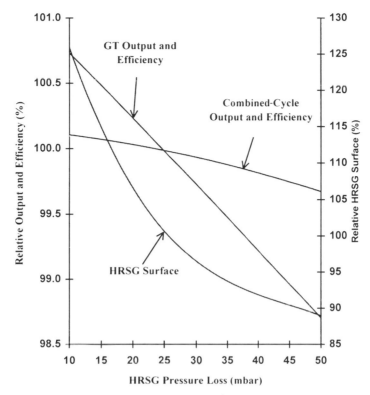

Figure 4-11 Influence of HRSG Back Pressure on Combined-Cycle Output and Efficiency, GT Output and Efficiency and HRSG Surface

feedwater temperature. A conventional steam plant attains a better efficiency if the temperature of the feedwater is brought to a high level by means of multi-stage preheating. There are two reasons for this difference.

First, a conventional steam generator is usually equipped with a regenerative air preheater that can further utilize the energy remaining in the flue gases after the economizer. This is not the case in an HRSG, where the energy remaining in the exhaust gases after the economizer is lost. In principle air coming into the gas turbine could be similarly preheated, further lowering the stack temperature, but this would lower the plant output significantly due to the decrease in air density at the compressor inlet, and the fact that gas turbines are volumetric machines so the inlet air flow to the gas turbine is always constant.

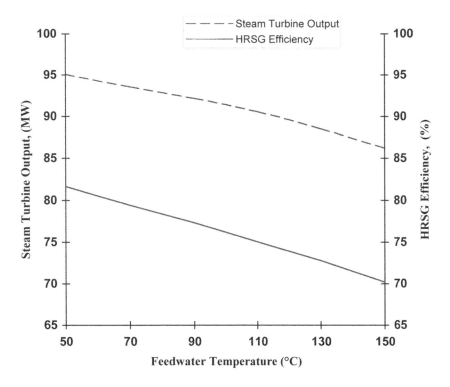

Figure 4-12 Effect of Feedwater Temperature on Steam Turbine Output and HRSG Efficiency for Cycles with One Stage of Preheating

Secondly (as shown in Fig. 4-13), the smaller temperature difference between the water and exhaust gases in the economizer of an HRSG is on the warmer end of the heat exchanger. That means the amount of steam production possible does not depend on the feedwater temperature. In a conventional steam generator, on the other hand, the smaller temperature difference is on the cold end of the economizer because the water flow is far larger in proportion to the flue gas flow. As a result, in conventional boilers the amount of steam production possible depends on the feedwater temperature.

Figure 4-14 shows two examples of conventional steam generators with different feedwater temperatures. It is obvious that with the same difference in temperature at the end of the economizer, the heat available for evaporation and superheating is significantly greater where the feedwater temperature is higher. Therefore, raising the feedwater temperature can increase the amount of live-steam produced by

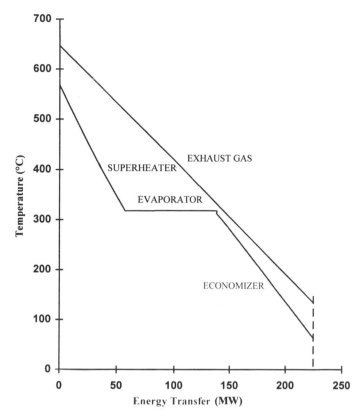

Figure 4-13 Energy/Temperature Diagram for a Single Pressure HRSG

a conventional boiler. Further, by preheating the boiler feedwater the water/steam cycle efficiency is increased due to the fact that more of steam is fully utilized and less energy is dissipated in the condenser.

In the single-pressure cycle example (Fig. 4-4), there is a single stage of preheating in the feedwater tank/deaerator and the feedwater temperature is relatively low, 60°C (140°F). However, stack temperature is 133°C (271°F) and contains energy which could be used to pre-heat the feedwater. One way to do this is by increasing the economizer surface and recirculating some of the heated water back to the feed-water tank to preheat the condensate. On entering the feedwater tank, the heated water undergoes a sudden pressure drop, causing it to flash into steam, which enables both preheating and deaeration to take place. As it is no longer necessary to extract the steam from the steam turbine

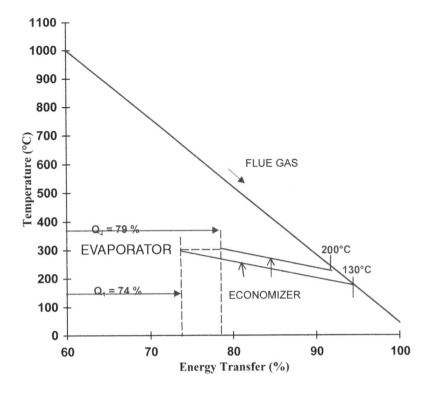

Figure 4-14 Energy/Temperature Diagram for a Conventional Steam Generator

for preheating, there is an increase in the steam turbine output. If all of the preheating in the example is done by recirculation, the gross power output of the cycle increases by 630 kW and the gross efficiency from 57.7% to 57.8%. The stack temperature falls from 133 to 112°C (271 to 234°F), demonstrating better use of the exhaust gas energy. The negative aspect of this is a major increase in the HRSG surface with a corresponding increase in HRSG cost.

The larger volume flow of exhaust steam produced could also lead to increased costs in the steam turbine and the condenser because their dimensions must be increased to accommodate it. The increase in the heat load of the condenser is more than the proportional increase for the gain in power output. The energy utilization rate of the HRSG rises by about 4% while the power output from the steam turbine increases only by 0.6%, because the additional exhaust heat recovered is at a low temperature level. The rate for converting it into mechanical energy is therefore low. The economical parameters of the plant must be used to determine whether this additional investment is worthwhile for the additional power output and efficiency.

This solution with economizer recirculation is only possible because the feedwater temperature is relatively low and the required preheating energy is available in the exhaust gas behind the evaporator. This may not be possible if the feedwater temperature has to be raised to avoid low temperature corrosion in the LP economizer. This is a chemical limitation on the energetic use of the exhaust gas. The corrosion, caused by water vapor and sulfuric acid in the exhaust gas, occurs whenever the gas is cooled below the acid dew point of these vapors.

In an HRSG, the heat transfer on the exhaust-gas side is inferior to that on the water or steam side. For that reason, the surface temperature of the pipes on the exhaust gas side is approximately the same as the water or steam temperature. If these pipes are to be protected against an attack of low-temperature corrosion, feedwater temperatures must remain approximately as high as the acid dew point. Thus, a high stack temperature for the exhaust gases does no good if the temperature of the feedwater is too low. Low-temperature corrosion can occur even when burning fuels containing no sulfur if the temperature drops below the water dew point. This is generally between 40 and 45°C (104 and 113°F).

The sulfuric acid dew point temperature for a fuel depends on the quantity of sulfur in that fuel. A feedwater temperature of 60°C (140°F) as in our example corresponds to a gas fuel with low sulfur content (<3ppm sulfur in fuel). Oil tends to have more sulfur, resulting in feedwater temperatures of around 120 to 160°C (248 to 320°F). Here multi-stage preheating, as in a conventional cycle, would improve the efficiency but generally single-pressure cycles are chosen where efficiency is not highly evaluated and further investment would not be economically viable.

Single-Pressure Cycle with a Preheating Loop in the HRSG

For cycles burning fuels with a high sulfur content, the easiest improvement to the single-pressure cycle is to use an additional heat exchanger at the end of the HRSG to recover heat for feedwater preheating. This preheating loop must be designed so that temperatures do not drop below the acid dew point. It is therefore normal practice to install an evaporator loop in the HRSG operated at a pressure equivalent to the acid dewpoint temperature.

Figure 4-15 shows a solution, in which a low-pressure evaporator (6, 8) generates saturated steam solely for the feedwater tank/deaerator. In this case, because this loop is at a low pressure at the low-temperature end of the HRSG, the power required to drive the additional feedwater pump (15) is quite small.

The deaerator/feedwater tank could alternatively be integrated into the drum of the preheating loop, which then functions as a feedwater tank for the cycle in providing a water buffer. This simplifies the

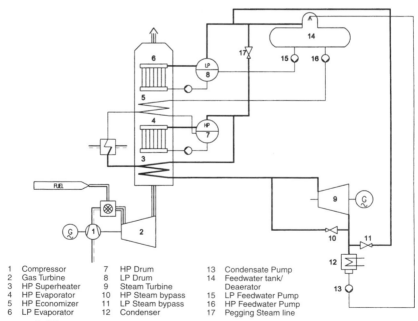

1	Compressor	7	HP Drum	13	Condensate Pump
2	Gas Turbine	8	LP Drum	14	Feedwater tank/
3	HP Superheater	9	Steam Turbine		Deaerator
4	HP Evaporator	10	HP Steam bypass	15	LP Feedwater Pump
5	HP Economizer	11	LP Steam bypass	16	HP Feedwater Pump
6	LP Evaporator	12	Condenser	17	Pegging Steam line

Figure 4-15 Flow Diagram of a Single Pressure Cycle with LP Preheating Loop for High-Sulphur Fuels

cycle because additional feedwater pumps and level controls are not required.

Figure 4-16 shows the required and available energy in a pre-heating loop as a function of feedwater temperature and live-steam pressure. The required energy is that needed to raise the temperature of the full condensate flow from condenser saturation temperature to the feedwater temperature. The available energy is that in the exhaust gases downstream of the economizer. An eventual stack temperature of 12 K (22°R) above the feedwater temperature–independent of the live-steam pressure–is assumed. The higher the feedwater temperature, the more preheating is required. The available energy depends on the live-steam pressure for the following reasons. The exhaust gas temperature after the economizer of the HRSG rises as the live-steam pressure rises because less steam is produced at higher pressures so less energy is removed from the economizers, leaving more energy available for pre-heating. A lower feedwater temperature also results in more available

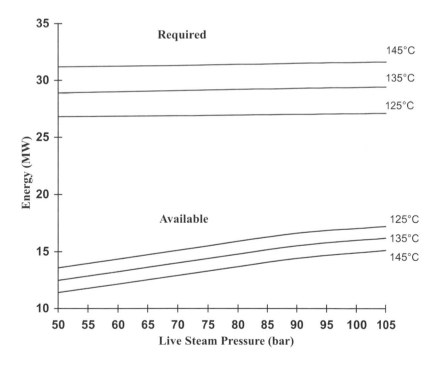

Figure 4-16 Effect of Live Steam Pressure and Feedwater Temperature on Available Heat Compared to Required Heat in Preheating Loop

energy because the exhaust gases can be cooled to a lower level even though more heating is required. In this case at about 100°C (212°F) the available energy matches required energy.

Figure 4-16 clearly demonstrates that at these feedwater temperatures there is not enough energy in the HRSG to accommodate all of the necessary preheating of the feedwater. Therefore, some of the preheating must be done by other means. If a gas turbine with a lower exhaust gas temperature had been used, more energy would have been available for the preheating loop.

Figure 4-17 is an example of a cycle with an evaporator preheating loop and one additional LP condensate preheater fed by a steam turbine extraction of 7.1 kg/s (56,300 lb/hr) used to preheat the condensate to 80°C (176°F). A low-pressure preheater improves the steam process by extracting the steam at a lower level than if a steam turbine extraction to the feedwater tankhad been used. The additional preheating loop and preheater increase the plant cost but this investment is balanced by the improvement in efficiency. Even if the fuel contains very high levels of sulfur, the feedwater can, with this concept, be preheated to a sufficiently high temperature with a minor reduction in efficiency.

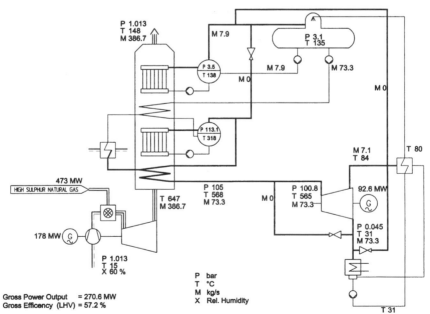

Figure 4-17 Heat Balance for a Single Pressure Cycle with LP Preheating Loop

Figure 4-18 shows the energy/temperature diagram for the HRSG with preheating loop. The exhaust gases are cooled by an additional 41 K (74°R) in the preheating loop. Despite these improvements the cycle still has an efficiency lower than that of the single-pressure cycle example (57.2% compared to 57.7%), because steam is extracted from the steam turbine to accommodate the higher feedwater temperature, (135°C (275°F) instead of 60°C (140°F)).

At a given feedwater temperature a single-pressure cycle provides better exhaust gas utilization when it has a preheating loop, as shown in Figure 4-18 compared to the single-pressure cycle. Nevertheless, that utilization is neither energetically nor exergetically optimum. The low-pressure evaporator could, at no great expense, produce more steam than required to preheat the feedwater, if feedwater preheating were done before this evaporator. That excess steam could be converted into mechanical energy if it were admitted into the steam

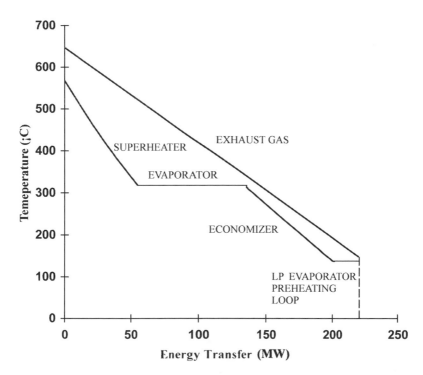

Figure 4-18 Energy/Temperature Diagram for a Single Pressure HRSG with LP Evaporator Preheating Loop

turbine at a suitable point. To do this, the steam turbine must have two steam admissions—one for high- pressure steam and another for low-pressure steam, and this would constitute a dual-pressure water/steam cycle.

Dual-Pressure Cycles

Dual-pressure cycles for high-sulfur fuels

A dual-pressure cycle for a fuel with a high sulfur content is shown in Figure 4-19. The LP evaporator loop (6, 8) generates steam for the steam turbine and for feedwater preheating in the feedwater tank (14). There are two stages of condensate preheating with low pressure preheaters (17) which are fed by extractions from the steam turbine. This makes good thermodynamic sense because the steam used to preheat the feedwater is of a low quality. As opposed to the sin-

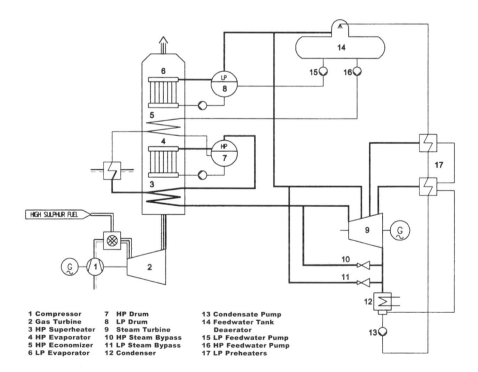

1 Compressor	7 HP Drum	13 Condensate Pump
2 Gas Turbine	8 LP Drum	14 Feedwater Tank
3 HP Superheater	9 Steam Turbine	Deaerator
4 HP Evaporator	10 HP Steam Bypass	15 LP Feedwater Pump
5 HP Economizer	11 LP Steam Bypass	16 HP Feedwater Pump
6 LP Evaporator	12 Condenser	17 LP Preheaters

Figure 4-19 Flow Diagram of a Dual Pressure Cycle for High Sulfur Fuels

gle-pressure cycle with a preheating loop, the excess LP steam is expanded in the steam turbine.

Figure 4-20 shows the effect on the steam turbine output of the number of preheaters for a range of feedwater temperatures. The higher the required feedwater temperature the lower the steam turbine output in all cases because less energy is available in the HRSG and more steam is required for preheating. Increasing the number of preheating stages increases the steam turbine output because the steam is used more efficiently. In order to raise the feedwater temperature to a certain level (e.g., 120°C (248°F)), steam must be extracted at a certain level in the steam turbine (say at a pressure corresponding to 130°C (266°F)). If this is done in one stage, all of the steam must be extracted at this pressure. If two stages are used, then a second extraction could be done at a pressure corresponding to say, 100°C (212°F), enabling this steam to expand slightly further in the steam turbine before being extracted, raising the steam turbine output. Although slightly more steam must be extracted at the lower level the effect of this on perfor-

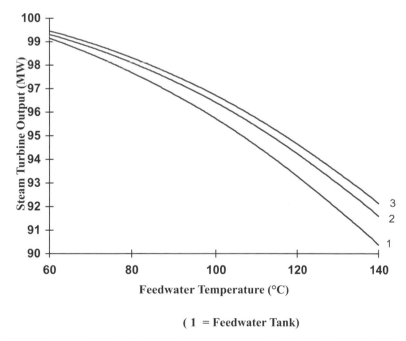

(1 = Feedwater Tank)

Figure 4-20 Effect of Feedwater Temperature and Number of Preheating Stages on Steam Turbine Output of a Dual Pressure Cycle

mance is negligible. A similar improvement is achieved with further stages of preheating.

Dual-pressure cycles for low sulfur fuels

If a fuel with low-sulfur content is being used and the feedwater temperature can be reduced, the LP feedwater preheaters of Figure 4-19 can be replaced by additional economizers in the HRSG. This enables more exhaust gas energy to be utilized, lowering the stack temperature. This is the most common type of dual-pressure cycle (Fig. 4-21). The first section in the HRSG (7) is a dual-pressure economizer, divided into an LP part for the LP feedwater and an HP part for the first step in heating the HP feedwater.

More of the steam generated in the LP evaporator and superheater (6, 5) is available for the dual-admission steam turbine. An extraction is made from the steam turbine for feedwater preheating and deaeration. Pegging steam (18) for off-design conditions for which the steam turbine extraction pressure is too low is taken from the LP livesteam line.

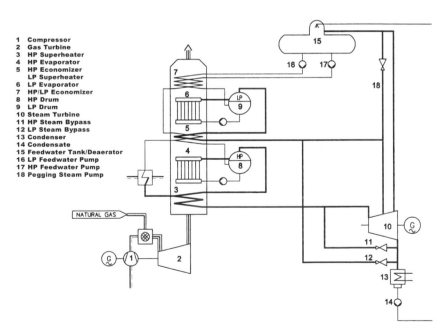

1 Compressor
2 Gas Turbine
3 HP Superheater
4 HP Evaporator
5 HP Economizer
   LP Superheater
6 LP Evaporator
7 HP/LP Economizer
8 HP Drum
9 LP Drum
10 Steam Turbine
11 HP Steam Bypass
12 LP Steam Bypass
13 Condenser
14 Condensate
15 Feedwater Tank/Deaerator
16 LP Feedwater Pump
17 HP Feedwater Pump
18 Pegging Steam Pump

NATURAL GAS

Figure 4-21 Flow Diagram of a Dual Pressure Cycle for Low Sulfur Fuels

Figure 4-22 shows the heat balance for a dual-pressure cycle burning natural gas with a low-sulfur content and a feedwater temperature of 60°C (140°F). Comparing this with the single-pressure cycle example (Fig. 4-4) shows clearly how the dual-pressure cycle makes better use of the exhaust gas in the HRSG resulting in a higher steam turbine output. The economizers at the end of the HRSG utilize the exhaust gases to a greater extent, lowering the stack temperature to 96°C (205°F) compared to 133°C (272°F) for the single-pressure cycle. The HP part of the HRSG is not affected by the presence of the LP section and HP steam production is the same. Overall steam production is increased due to the 5.4 kg/s (42,000 lb/hr) of LP steam. The gross efficiency of the cycle has risen from 57.7% to 58.6%–a considerable increase.

Figure 4-23 shows the heat flow diagram for the dual-pressure cycle. Compared to the single-pressure cycle (Fig. 4-5), the stack losses are reduced from 11.4% to 8.2% of the total fuel energy input.

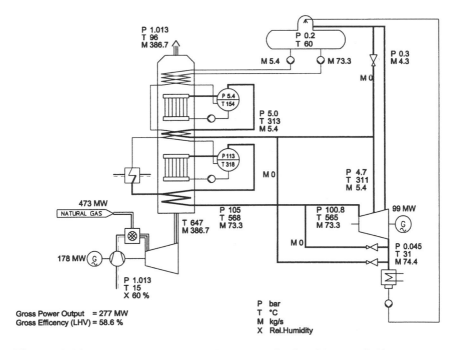

Figure 4-22 Heat Balance for a Dual Pressure Cycle with Low Sulfur Fuel

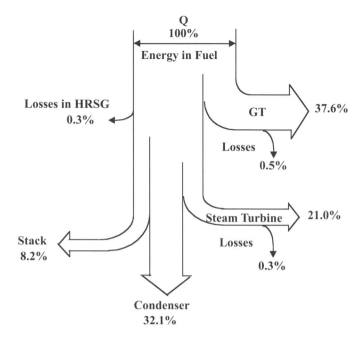

Figure 4-23 Energy Flow Diagram for the Dual Pressure Combined-Cycle Plant

However due to the increased LP steam flow more energy is lost in the condenser–32.1% instead of 29.9%. The portion of the steam turbine output has increased by 0.9%. This shows that the additional absorbed heat in the HRSG is converted with a moderate efficiency of 28% because it is being done at a low temperature level.

Figure 4-24 is the energy/temperature diagram for the dual-pressure HRSG. As it shows, most of the heat exchange takes place in the HP portion of the HRSG. This is directly related to the chosen pressure levels of the cycle. Comparing Figure 4-24 with Figure 4-13 shows how energy utilization at the cold end of the HRSG has been improved in the dual-pressure cycle.

Types of Deaeration

Deaeration is the removal of incondensable gases from the water or steam in the water/steam cycle. It is very important because a high oxygen content can cause corrosion of the components and pip-

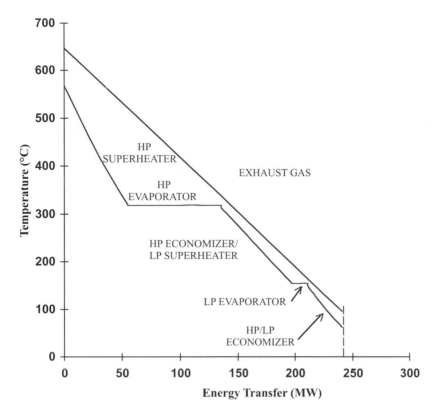

Figure 4-24 Energy/Temperature Diagram for a Dual Pressure HRSG

ing which come into contact with the medium. Typically an oxygen content of around 7 to 10 parts per billion (ppb) in the feedwater is recommended. Deaeration must be done continually because small leakages of air at flanges and pump seals in the part of the cycle under vacuum cannot be avoided. Indeed, in the design of these components, minimizing of leakage is an important factor.

Deaeration takes place when water is sprayed and heated, thereby releasing the absorbed gases. Generally, the condensate is sprayed into the top of the deaerator, which is normally placed on top of the feedwater tank. Heating steam, fed into the lower part, rises and heats the water droplets to the saturation temperature, releasing incondensable gases that are carried to the top of the deaerator and evacuated. The feedwater tank is filled with saturated and deaerated water that has steam hanging above it preventing any reabsorbtion of the air. Depending on the pressure in the feedwater tank/deaerator the deaeration process either takes place under vacuum (known as vacuum deaeration) or at a pressure above atmospheric pressure (known as over-pressure deaeration). Both concepts enable good deaeration to take place .

The feedwater system is simpler with vacuum deaeration than with overpressure deaeration because all of the feedwater heating can be done in the feedwater tank by adding the appropriate amount of heating steam without the need for additional heat exchangers. The heating steam required for vacuum deaeration is of a lower quality than that needed for overpressure deaeration, leaving more steam in the steam cycle for expansion in the turbine, thereby increasing turbine output and combined-cycle efficiency. Also, the feedwater temperature can be varied for the same system, for example, to accommodate oil operation.

However, with overpressure deaeration, as opposed to vacuum deaeration, the incondensable gases can be exhausted directly to atmosphere independently of the condenser evacuation system.

Both these single- and dual-pressure examples have been set up with vacuum deaeration. Feedwater temperature is 60°C (140°F), corresponding to a pressure in the feedwater tank/deaerator of 0.2 bar (2.9 psia). In contrast, the example with the preheaters has overpressure deaeration at 135°C, 3.1 bar (275°F, 30 psig).

Deaeration also takes place in the condenser. The process is similar to that in the deaerator. The turbine exhaust steam condenses and collects in the condenser hotwell while the incondensable gases are extracted by means of evacuation equipment. Again, a steam cushion separates air and water so reabsorbtion of the air cannot take place. Levels of deaeration in the condenser often can be achieved which are comparable to those in the deaerator. Sometimes, therefore, the separate deaerator/feedwater tank can be eliminated from the cycle and the condensate fed directly to the HRSG from the condenser. For cycles of this type the condenser hotwell capacity must be increased to provide the cycle with a water buffer because there is no feedwater tank. To raise the condensate temperature above the sulfur dew point temperature before it enters the HRSG feedwater is recirculated from the HRSG to a point upstream of the economizer inlet.

An important factor concerning deaeration is the quantity of makeup water and where this is to be admitted to the cycle. Makeup water is usually fully saturated with oxygen, but if the required quantity is less than about 25% of the steam turbine exhaust flow, condenser deaeration may be appropriate with the makeup water being sprayed into the cycle in the condenser neck. For cycles with process extractions and therefore high makeup water quantities a separate deaerator is normally preferred.

Main Design Parameters of the Dual-Pressure Cycle

Live-steam pressure

Two things must be noted when selecting the HP and LP live-steam pressures for dual-pressure cycles.

First, the HP steam pressure must be relatively high to attain good exergetic utilization of the exhaust gas; and, second, the LP steam pressure must be low to attain good energetic and exergetic utilization of the exhaust gas heat and hence to achieve a higher steam turbine output.

Figure 4-25 shows the steam turbine output for the dual-pressure cycle as a function of the HP steam pressure for various LP steam pressures. Other parameters remain constant. As either pressure rises the steam turbine power output increases for a given steam mass flow, due to the increased enthalpy drop in the steam turbine. However, at higher pressures less steam is generated so there is a trade-off between the higher enthalpy drop and lower mass flow. This is more pronounced for the LP section due to the relatively big influence that a change in pressure will have.

The pressure in the LP evaporator should not be below about 3 bar (29 psig) because the enthalpy drop available in the turbine becomes very small and the volume flow of steam becomes very large and the hardware is therefore more expensive. LP pressure is chosen at 5 bar (58 psig). The moisture content also rises with the HP pressure. This is why the HP pressure is not increased beyond 105 bar (1508 psig) for an LP pressure of 5 bar (58 psig) as in the example.

Figure 4-26 shows how HRSG efficiency improves as the LP pressure decreases. However, for thermodynamic reasons, the best en-

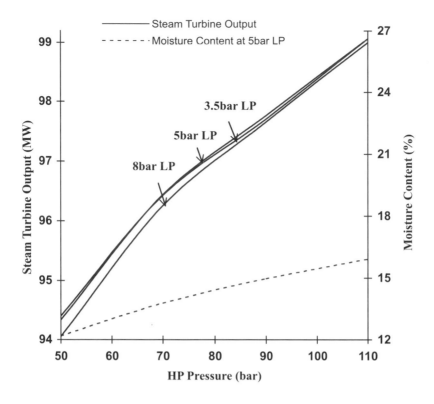

Figure 4-25 Effect of HP and LP Pressure on Steam Turbine Output and Exhaust Moisture Content for a Dual Pressure Cycle

ergy utilization of the HRSG does not necessarily result in the highest steam turbine output as was seen for the single-pressure cycle (Fig. 4-9).

Live-steam temperature

Unlike the single-pressure cycle, in the dual-pressure cycle a live-steam temperature increase brings with it a substantial improvement in the output. The same arguments made for the single-pressure cycle are valid for the HP section where a slight decrease in output for an increase in temperature was seen. However in the dual-pressure cycle, more energy is made available to the LP section if the HP temperature is raised, which more than compensates for the slight output loss of the HP steam.

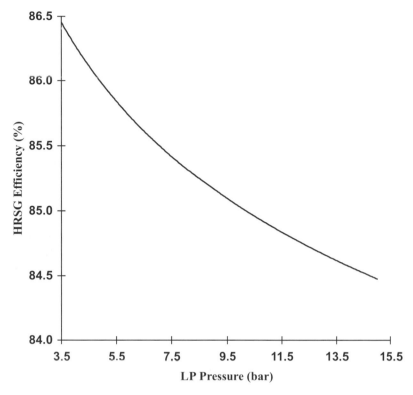

Figure 4-26 Effect of LP Pressure on HRSG Efficiency for a Dual Pressure Cycle

In Figure 4–27, the extent of this improvement on steam turbine output is shown. In the LP section as opposed to the HP section, a high degree of superheat improves the output marginally. The curve was calculated with a constant pressure drop in the LP superheater. In reality, this pressure drop would be in proportion to the degree of superheat raising the curve slightly towards point A. A cycle without a superheater could be advantageous for the mechanical design of the system. The benefit of increasing the HP temperature is significant for steam turbine output. Increasing the LP temperature is important because it leads to a reduction in the moisture content in the last stage of the steam turbine. There is also a reduction in the HRSG surface because although the LP superheater surface increases, those of the LP evaporator and economizer will decrease due to the smaller mass flow, and the net result is a decrease.

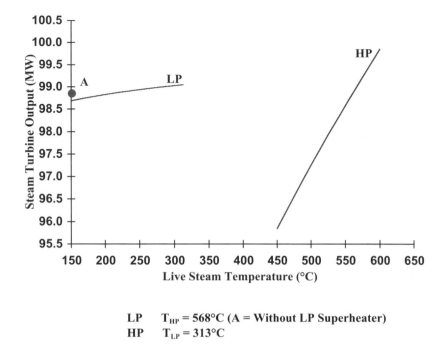

LP T_{HP} = 568°C (A = Without LP Superheater)
HP T_{LP} = 313°C

Figure 4-27 Effect of LP and HP Steam Temperatures on Steam Turbine Output for a Dual Pressure Cycle

Pinch point

The pinch-point of the HP evaporator has less influence here than in a single-pressure cycle because the energy that is not utilized in the HP section can be recovered in the LP section of the HRSG. Transferring energy from the HP to the LP section by increasing the HP pinch-point causes exergetic losses because HP steam is more valuable than LP steam due to the greater enthalpy drop in the steam turbine. Figure 4-28 shows the steam turbine output and relative HRSG surface as a function of the pinch-points of the HP and LP evaporators.

The HP and LP pinch-points are interrelated. A reduction of the HP pinch-point increases the surface of the HP evaporator and economizer but also reduces that of the LP section. This is because the exhaust gas temperature after the HP economizer falls, reducing the amount of heat available for the LP section and therefore the LP steam

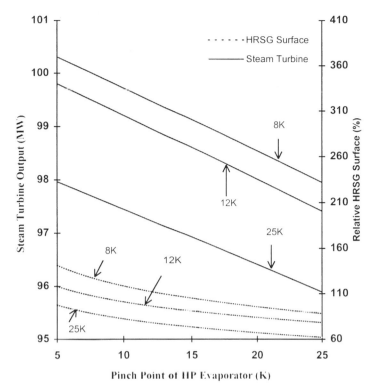

Figure 4-28 Effect of HP and LP Evaporator Pinch Point on Steam Turbine Output and Relative HRSG Surface for a Dual Pressure Cycle

flow. The heat required for LP feedwater heating hence decreases since the LP feedwater flow is smaller.

For a dual-pressure cycle, the pinch-points of both the HP and the LP evaporators have less effect on the output of the steam turbine than the pinch-point in a single-pressure cycle.

Triple-Pressure Cycle

If a third pressure level is added to the dual-pressure cycle a further improvement can be made, mainly by recovering more exergy from the exhaust gas. In the flow diagram shown in Figure 4-29, separate pumps (20, 21) supply feedwater to a dual-pressure economizer (9), one at the HP and the other at the intermediate pressure (IP) level. On leaving the IP economizer, the IP feedwater divides into two

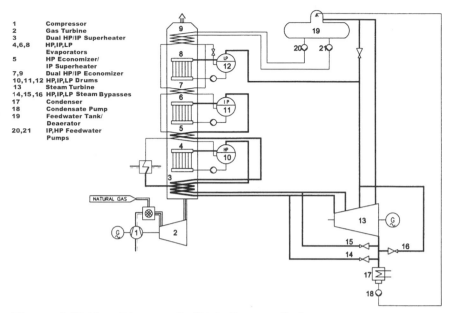

1	Compressor
2	Gas Turbine
3	Dual HP/IP Superheater
4,6,8	HP,IP,LP Evaporators
5	HP Economizer/ IP Superheater
7,9	Dual HP/IP Economizer
10,11,12	HP,IP,LP Drums
13	Steam Turbine
14,15,16	HP,IP,LP Steam Bypasses
17	Condenser
18	Condensate Pump
19	Feedwater Tank/ Deaerator
20,21	IP,HP Feedwater Pumps

Figure 4-29 Flow Diagram of a Triple Pressure Cycle

parts–one entering a second dual-pressure economizer (7) and the other being throttled into the LP steam drum (12). The saturated LP steam collecting in the drum is fed directly to the steam turbine. The HP and IP pressure levels follow the pattern of economizer, evaporator, and superheater until the superheated steam generated is fed to the triple-pressure steam turbine. Again, each pressure level has a separate steam turbine bypass (14, 15, 16).

Figure 4-30 shows the heat balance for a triple-pressure cycle with a low-sulfur natural gas fuel and a feedwater temperature of 60°C (140°F). For this example there is only a slight improvement in output over the dual-pressure cycle, with efficiency improving marginally from 58.6% to 58.7%. The stack temperature is the same. The improvement is due to a slight exergetic gain caused by the IP level reducing the area between the exhaust gas and the water/steam lines (as shown on the temperature/energy diagram, Fig. 4-31). Compared to the HP flow of 72.5 kg/s (575,200 lb/hr), the IP and LP flows are very small–only 3.1 kg/s (24,600 lb/hr) and 3.0 kg/s (23,800 lb/hr) respectively. This limits the potential gain in exergy due to the addition of the IP section. The HP flow is slightly less than that for the dual-pressure

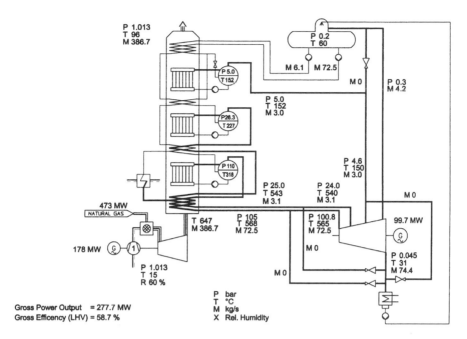

Figure 4-30 Heat Balance for a Triple Pressure Cycle

example because the IP superheater is at a high temperature level, removing energy from the HP section of the HRSG. This has a minor negative effect on output and is part of the reason why the benefit in this triple pressure cycle is so small. A benefit is gained because IP steam is generated in the place of some LP steam.

The energy flow diagram (Fig. 4-32) shows the slight improvement in the steam turbine output over the dual-pressure cycle for the reasons stated above. The stack losses are the same (same stack temperature) but there is a slight decrease in heat load in the condenser that corresponds to the gain in steam turbine output. For gas turbines with lower exhaust gas temperatures, more energy would be available for IP steam production because the HP steam production would be lower making this concept more attractive.

Main Design Parameters of the Triple-Pressure Cycle

Figure 4-33 shows the effect of the HP and IP pressure level on the steam turbine output for a constant LP pressure. These pressures

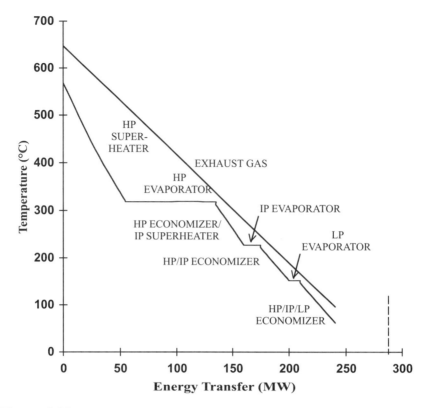

Figure 4-31 Energy/Temperature Diagram for a Triple Pressure HRSG

should be such that the best possible exergy utilization of the exhaust gas is achieved. In order to do this, the IP level must increase as the HP level increases. At low HP pressures an IP pressure of 10 bar (130 psig) brings more output than 30 bar (420 psig). If the HP pressure is increased to 105 bar (1508 psig) the 20 bar (275 psig) IP pressure gives the highest steam turbine output.

Moisture at the steam turbine exhaust plays a part and sets the HP limit to 105 bar (1508 psig). The optimum IP pressure at this point is 25 bar (248 psig) which has been used for the example.

For a given HP and IP pressure, the maximum steam turbine output is at a clearly definable LP pressure (Figure 4-34). As for the dual-pressure cycle, LP pressures below 3 bar (29 psig) are not recommended. The HRSG surface decreases as the LP pressure increases because less heat exchange takes place at the low temperature end of the

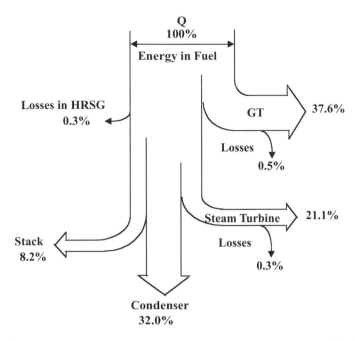

Figure 4-32 Energy Flow Diagram for the Triple Pressure Combined-Cycle Plant

HRSG, thus reducing the cost of the HRSG and other hardware. In the example, the LP pressure was chosen to be 5 bar (58 psig).

Figure 4-35 shows the relationship between the live-steam temperatures and steam turbine output. The HP steam temperature has a significant effect whereas higher IP and LP steam temperatures only slightly increase the power output. There is, however, another advantage with a high IP temperature.

By heating the IP steam to a temperature level close to that of the HP live-steam, a small reheat effect is seen on the steam turbine expansion line. This is because the hot IP steam mixes with the HP steam that has already expanded in the HP turbine. This decreases the moisture content and therefore the risk of erosion due to wetness in the last stage of the steam turbine. Figure 4-36 shows this effect on an enthalpy/entropy diagram for steam. The so called "mild reheat" causes the steam turbine expansion line to move to the right where wetness is lower. However, there is no improvement in moisture content in this triple pressure cycle compared with the dual pressure cycle because

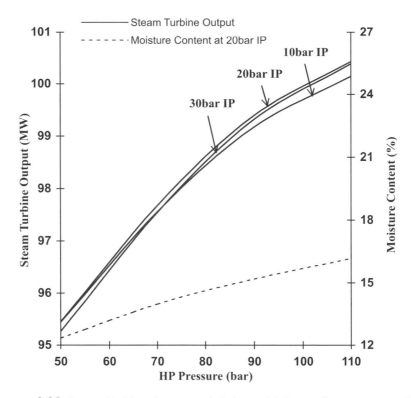

Figure 4-33 Steam Turbine Output and Exhaust Moisture Content versus HP and IP Pressure for Triple Pressure Cycle at Constant LP Pressure (5 bar)

any benefit derived from the IP superheater is cancelled out by the elimination of the LP superheater.

Care must be taken to limit the thermal stresses within the steam turbine arising from a temperature difference between the HP steam after expansion and the IP live-steam at the mixing point. This is solved by the constructional design of the steam turbine casing at the IP steam admission point.

For a modern gas turbine (like the one used in the example), the potential gain due to this mild reheat effect is limited because of the low quantity of IP steam (approximately 5% of the HP steam flow). This limits the temperature increase at the mixing point in the steam turbine. A similar mild reheat effect would be gained if there were a su-

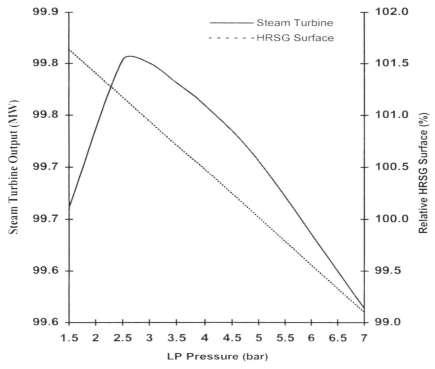

Figure 4-34 Effect of LP Pressure on Steam Turbine Output and Relative HRSG Surface for Triple Pressure Cycle at Constant HP (105 bar) and IP (25 bar)

perheater in the LP section. However, this would make the cycle more complex with practically no gain in performance.

Figure 4-37 shows the effect on steam turbine output and HRSG surface when varying the HP and IP pinch-points with the LP pinch-point held constant. The curve is similar to that of Figure 4-28 for the dual-pressure cycle. Steam turbine output increases as either pinch-point is reduced and the HRSG surface increases, exponentially as the pinch-points tend to zero. The steam turbine output is actually *lower* in the triple-pressure cycle than it was in the dual-pressure cycle example at this point. This is because the dual-pressure cycle has an LP superheater that compensates for this effect. If the dual-pressure cycle had had no LP superheater this would not be the case.

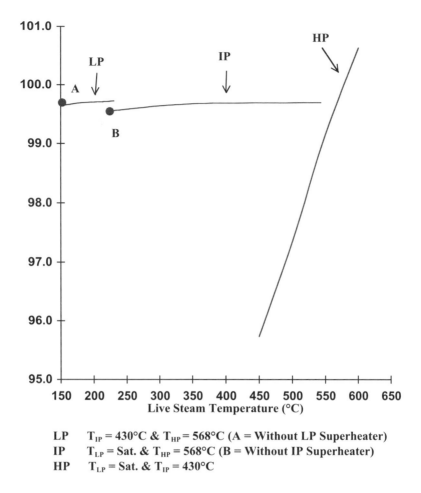

Figure 4-35 Live Steam Temperature Optimization for a Triple Pressure Cycle

Reheat Cycles

From our analysis thus far, it is clear that the moisture content in the steam turbine exhaust is significant in limiting further improvements in the performance of the various cycles. By extending the idea of mild reheat, the full reheat cycle is derived in which this moisture content is reduced and improvement in the performance is possible. The idea is that (taking the triple-pressure cycle as an example) after the expansion of the HP steam in the steam turbine to IP level, the

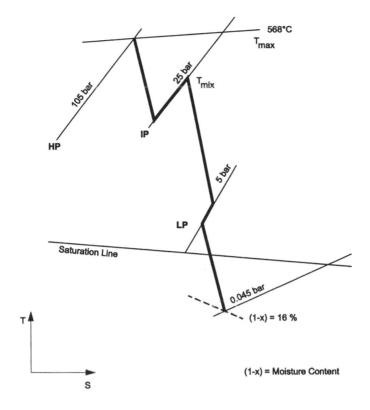

Figure 4-36 Temperature/Entropy Diagram Showing the Effect of "Mild Reheat" on the Steam Turbine Expansion Line

steam returns to the HRSG and mixes with the steam leaving the IP superheater. This steam is then heated to a temperature similar or equal to the HP live-steam temperature before being admitted to the steam turbine. In a dual-pressure reheat cycle there is no mixing on re-entering the HRSG, the cold reheat goes to an independent reheater section.

The reheater takes more exergy out of the HRSG and increases the steam turbine enthalpy drop resulting in a higher steam turbine output. Because the last part of the steam turbine expansion line is moved further to the right on the enthalpy/entropy diagram, there is a reduction in the moisture content.

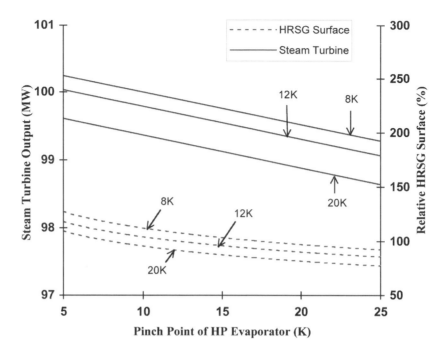

Figure 4-37 Effect of HP and IP Evaporator Pinch Points on Steam Turbine Output and Relative HRSG Surface for a Triple Pressure Cycle with Constant LP Pinch Point

Triple-pressure reheat cycle

Figure 4-38 is the flow diagram of the triple-pressure cycle with reheat. The steam turbine has separate HP and IP/ LP casings (13, 14) in order to accommodate the extraction of the cold reheat steam. The high pressure bypass (15), instead of dumping steam into the condenser, dumps it into the cold reheat line (i.e., the line leaving the HP steam turbine).

The heat balance is shown in Figure 4-39. The HP live-steam pressure–now increased to 120 bar (1,725 psig)–together with the reheater (which removes energy at the hot end of the HRSG) leads to a reduced HP mass flow of 59.2 kg/s (469,700 lb/hr) compared to the triple-pressure cycle. The IP steam flow is, for that reason, slightly higher (5.9 kg/s (46,800 lb/hr)). However, there is a significant improvement in the cycle performance because, due to the reheat, there is a greater exergy transfer in the hot end of the HRSG. The IP steam

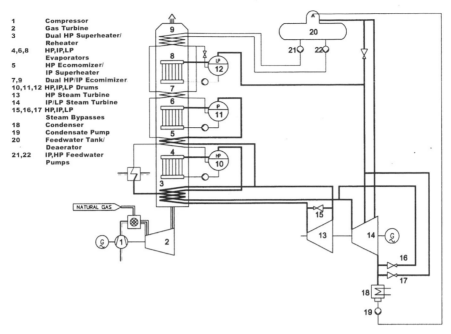

Figure 4-38 Flow Diagram of a Triple Pressure Reheat Cycle

is expanded from a high temperature level 65.1 kg/s (516,500 lb/hr) from 565°C (1049°F) instead of 75.6 kg/s (599,800 lb/hr) (HP and IP steam) from the mix temperature of 354°C (669°F) in the triple-pressure cycle. The resulting gross output is 2.8 MW higher than for the triple-pressure cycle with an efficiency of 59.3% instead of 58.7%.

The decrease in the moisture content is illustrated on the enthalpy/entropy diagram (Fig. 4-40). The value is 10% compared to 16% in the triple-pressure example. The percentage of the stack losses has increased 0.4% (Fig. 4-41) because the stack temperature is slightly higher 103°C, (217°F) . This occurs when the reheater "steals" energy from the HP section of the HRSG, resulting in less HP steam production and therefore less feedwater heating in the HP economizers. Increases in IP and LP mass flows do not compensate for this and so the stack temperature increases. Combined with the additional steam turbine output, this explains the lower heat loads in the condenser.

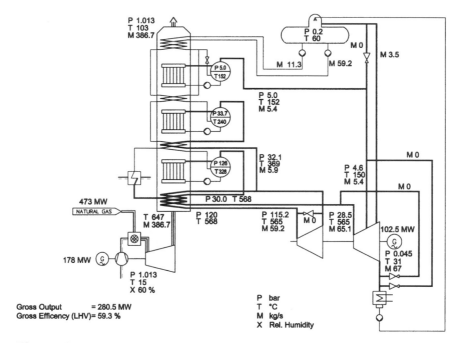

Figure 4-39 Heat Balance for a Triple Pressure Reheat Cycle

The energy/temperature diagram for the triple pressure reheat cycle is shown in Figure 4-42 The energy taken out in the HP super-heater/reheater, and HP evaporator is approximately 138 MW, compared to 140 MW for the triple-pressure cycle. This is mainly due to the HP evaporator pressure level difference. However, the energy is transferred at a higher temperature level and because the combined mass flow of the HP superheater/reheater is higher than that of the HP/IP superheater in the triple pressure cycle, this part of the diagram moves closer to the exhaust gas temperature. This results in an exergy gain and therefore a higher steam turbine output.

Live-steam data

The effect of live-steam and reheat pressure on the steam turbine output is shown in Figure 4-43. Power increases with live-steam pressure although the improvement is less after a certain point depending on the reheat temperature. The IP pressure is chosen to match

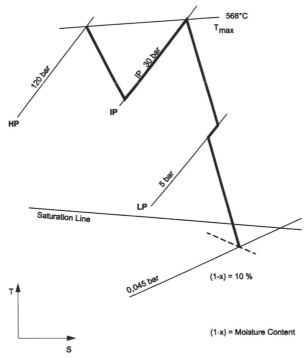

Figure 4-40 Temperature/Entropy Diagram Showing the Effect of Full Reheat on the Steam Turbine Expansion Line

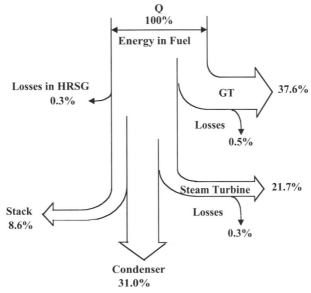

Figure 4-41 Energy Flow Diagram for the Triple Pressure Reheat Combined-Cycle Plant

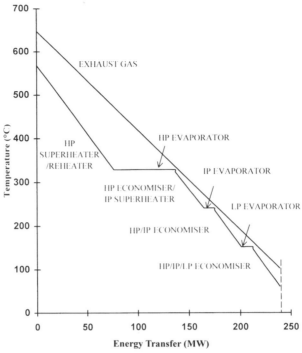

Figure 4-42 Energy/Temperature Diagram for a Triple Pressure Reheat HRSG

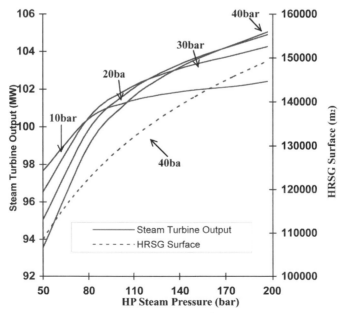

Figure 4-43 Steam Turbine Output and HRSG Surface versus HP and Reheat Pressure for a Triple Pressure Reheat Cycle at Constant HP (568°C) and Reheat (568°C) Temperature

the reheat pressure after accounting for losses in the reheater and cold reheat piping.

An increase in the HP pressure leads to an increase in the cost of the HRSG, feedwater pumps, and piping which is often the limiting factor when determining the correct pressure. There will also be an increase in the auxiliary consumption of the feedwater pumps as the HP pressure rises which means that the benefit in the net power output is slightly less than shown on the curve.

At low live-steam pressures a lower reheat pressure is advantageous because the enthalpy drop in the HP steam turbine is greater before reheating. The two pressures should not be too close together (as illustrated by the point with 50 bar live-steam pressure and 40 bar reheat pressure). Similarly as the HP pressure increases a higher reheat pressure is preferable.

It is important to keep the steam-side pressure loss in the cold reheat piping, reheater, and hot reheat piping as low as possible. Increasing the pressure loss has the same effect as throttling over the steam turbine valves. The steam turbine expansion line is moved towards the right on the enthalpy/entropy diagram, resulting in a higher exhaust enthalpy and therefore a lower steam turbine output.

Figure 4-44 shows the effect of the HP live-steam and reheat-steam temperature on the steam turbine output. An increase in the temperature increases the output significantly. The curve shows both temperatures being varied at the same time each contributing approximately half the gain.

The relationship between the pinch-points of this cycle are the same as those for the triple-pressure non-reheat cycle.

High Pressure Reheat Cycles

It is clear that there is a performance benefit in a reheat cycle because of the greater expansion in the steam turbine and the improved exergy utilization in the HRSG. Still more can be achieved if the live-steam pressure can be raised. State of the art technology is available with which this can be done. It has been shown that for dual and triple-pressure cycles, the HP part of the cycle makes the main contribution to the performance so, therefore the major investment should be placed here.

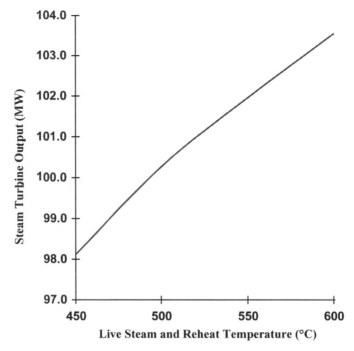

Figure 4-44 Steam Turbine Output versus HP Steam and Reheat Temperature for a Triple Pressure Reheat Cycle at Constant HP (120 bar), IP (30 bar) and LP (5 bar) Pressure

Figure 4-45 is the flow diagram of a cycle that has been derived with these factors in mind to achieve a higher efficiency at a lower level of complexity. It is a high-pressure reheat cycle and a dual-pressure cycle because the aim was to improve the exergy exchange in the HP part of the HRSG. To accommodate the higher HP steam pressure a "once-through" HRSG HP section (2,4) is chosen. Unlike the conventional drum system here there is no drum and the economizer, evaporator and superheater form a continous HRSG section. The separator (5) ensures safe operation and control of the HRSG. It will, for example, prevent water from entering the superheater, and provide a means of blowdown for the HP section. In normal operating conditions the separator is dry. The same equipment is used in conventional plants with once-through boilers. The LP part of the HRSG is of the drum type design. The increased HP pressure improves the exergy transfer in the HP section, although this improvement is partly reduced because there is no IP section which is omitted in order to reduce the complex-

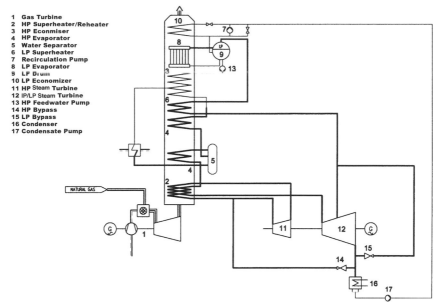

1 Gas Turbine
2 HP Superheater/Reheater
3 HP Econmiser
4 HP Evaporator
5 Water Separator
6 LP Superheater
7 Recirculation Pump
8 LP Evaporator
9 LP Drum
10 LP Economizer
11 HP Steam Turbine
12 IP/LP Steam Turbine
13 HP Feedwater Pump
14 HP Bypass
15 LP Bypass
16 Condenser
17 Condensate Pump

Figure 4-45 Flow Diagram of a High Pressure Reheat Cycle with a HP Once Through HRSG and a Drum Type LP Section

ity of the cycle. For steam turbine bypass operation the reheater is run dry and all HP steam is dumped into the condenser.

The cycle has no feedwater tank, so deaeration is done in the condenser and a recirculation loop in the LP economizer of the HRSG is used to raise the temperature of the incoming feedwater. The condensate pumps (17) must be designed to bring the condensate to the LP drum, fulfilling the function of LP feedwater pumps as well. The HP feedwater pumps (13) take suction from the LP drum.

The heat balance is shown in Figure 4-46. As expected, performance is further improved. The gross output is 2.4 MW above that of the triple-pressure reheat cycle and efficiency has risen from 59.3% to 59.8%–a considerable improvement. Part of this benefit comes because a once-through HRSG has by design an approach point temperature of zero but the same pinch-point (12 K (22°R)) is used here as in the other examples. However, due to the increased pressure of the HP system, power consumed in the feedwater pumps will be higher, slightly reducing the gain in the net power output. The example is offered with a fuel

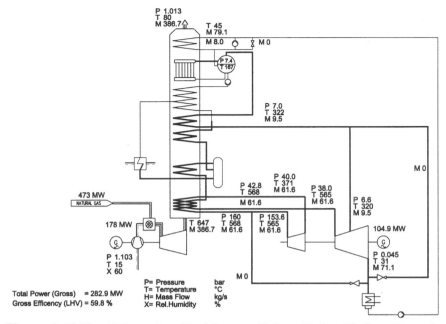

Figure 4-46 Heat Balance for a Dual Pressure Reheat Cycle with Once Through HRSG

that has no sulfur content so the feedwater temperature has been reduced to optimize the energy in the cold end of the HRSG. In such a cycle the feedwater temperature has an effect on the HRSG surface but not on the performance so long as there is sufficient energy available in the exhaust gas to preheat the incoming condensate.

Gross efficiencies approaching 60% are the current state of the art for combined-cycle power plants with this concept under the given ISO conditions.

Cycles with Supplementary Firing

Supplementary firing is a way of increasing the plant output by installing duct burners in the HRSG inlet duct that add energy to the cycle by increasing the exhaust gas temperature often at the expense of efficiency. Supplementary firing is appropriate for HRSGs because there is usually sufficient oxygen content in the exhaust gas to act as combustion air. In an open cycle gas turbine with a single stage of combustion, only 30-50% of the oxygen contained in the air is used for combustion.

Earlier combined-cycle installations generally had supplementary firing. This is not the case today due to progress in the development of the gas turbine. As gas-turbine inlet temperatures and hence exhaust gas temperatures increase, the importance of supplementary firing diminishes for two reasons. First, because the temperature window between the gas turbine exhaust and the duct burner exhaust decreases so the added benefit of supplementary firing decreases. Second, optimum values can be given to the water/steam cycle parameters with the gas turbine alone. Since current levels of efficiency are set by reheat cycles, the efficiencies achieved in cycles with supplementary firing lead to a high cost of electricity.

Nevertheless, increased operating and fuel flexibility of the combined-cycle with supplementary firing may be an advantage in special cases, particularly in installations used for cogeneration of heat and power where this arrangement makes it possible to control the electrical and thermal outputs separately (see chapter 5).

Figure 4-47 shows energy/temperature diagrams for a single-pressure HRSG with constant live-steam conditions and inlet exhaust gas temperatures of 647°C, (1,197°F), 750°C, (1,382 °F) and 1,000°C, (1,832°F), the latter two, after supplementary firing. At 647°C the temperatures of gas and water in the economizer are convergent, with the minimum difference in temperature on the evaporator end, typical for an HRSG without supplementary firing. At 1,000°C (1,832°F), on the other hand, the minimum difference in temperature is at the inlet to the economizer on the water side. This pattern corresponds more to that of a conventional steam generator.

A temperature of 750°C (1,382°F) after supplementary firing gives the best exergetic utilization of the exhaust gas with a constant difference in temperature along the entire economizer. The single pressure cycle is here at an optimum with the exhaust gas cooled down to a temperature close to the feedwater temperature. There is no exergy or energy available for additional pressure levels.

Figure 4-48 shows how relative power output and efficiency depend on the temperature after supplementary firing for a single- and a dual-pressure cycle. The reference point at 647°C (1,197°F) is the single-pressure cycle without supplementary firing. Above 750°C (1,382°F) there is no longer any performance benefit in the dual-pressure cycle. Calculations assume the use of natural gas fuel. When burn-

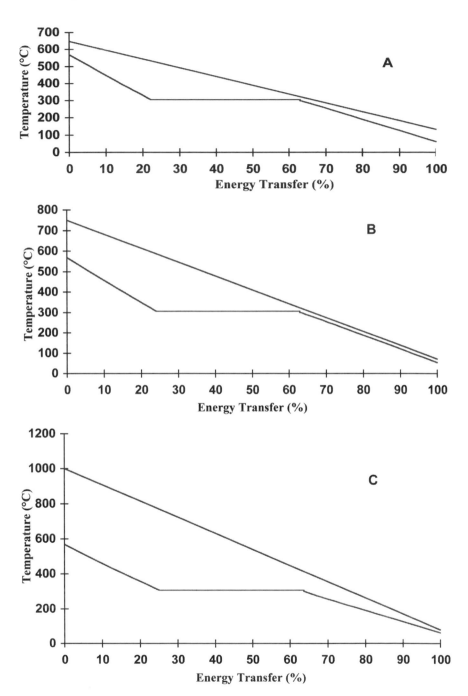

Figure 4-47 Energy/Temperature Diagram for 647°C(A), 750°C(B) and 1000°C(C) Exhaust Gas Temperature Entering the HRSG

ing oil, the paths of the curves for the single-pressure system would not be significantly changed, but there would be less difference between the single and dual-pressure systems. Triple-pressure and triple-pressure reheat cycles follow the same pattern, starting out with a larger difference without supplementary firing and ending at the same point at 750°C (1,382°F).

For installations with supplementary firing that are frequently operated at part loads, it could make economic sense to select a cycle with more pressure levels or even reheat. At part loads or when the supplementary firing is switched off, the exhaust gas temperature at the outlet of the HP section rises and an LP section would enable this exhaust gas energy to be used.

Figure 4-49 shows the heat balance of a typical combined-cycle plant with supplementary firing to 750°C, (1,382°F) with natural gas and a feedwater temperature of 60°C (140°F). The basic arrangement for this installation is the same as that for the single-pressure system in Figure 4-4.

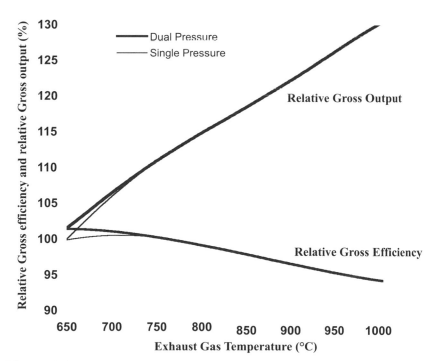

Figure 4-48 Effect of Temperature after Supplementary Firing on Power Output and Efficiency Relative to that of a Single Pressure Cycle

Supplementary firing increases the steam turbine output by 24.3 MW compared to the cycle without supplementary firing. The efficiency rises slightly from 57.7% to 57.9% because the increased steam production also results in increased mass flows through the economizers removing more energy from the exhaust gas, thus lowering the stack temperature. Generally, for cycles with more pressure levels supplementary firing has a negative effect on the efficiency because without supplementary firing these cycles already make maximum use of the exhaust gas energy.

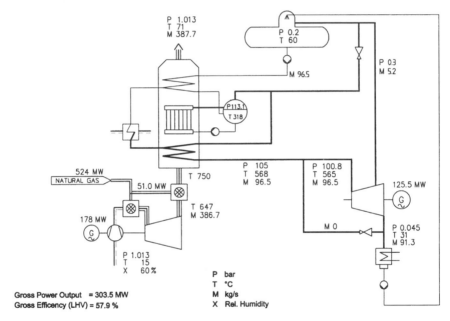

Figure 4-49 Heat Balance for a Single Pressure Cycle with Supplementary Firing

Summary of Cycle Performance

To give an overview of all the cycles discussed in this section, the performance data for the examples in this section are summarized in Table 4-1. The auxiliary power consumption for each case and the resulting net output and efficiency are also given, which are important for comparing the cycles in real terms.

Table 4-1 Performance Comparison for Different Cycle Concepts (Natural Gas Fuel with Low Sulfur Content)

		Single-pressure	Dual-pressure	Triple-pressure	Triple-pressure reheat	Dual-pressure reheat	Single-pressure/ supplementary firing
Gas Turbine Fuel Input (LHV)	MW	473	473	473	473	473	473
Duct Burner Fuel Input (LHV)	MW	0	0	0	0	0	51
Total Fuel Input (LHV)	MW	473	473	473	473	473	524
Gas Turbine Output	MW	178	178	178	178	178	178
Steam Turbine Output	MW	94.8	99.0	99.7	102.5	104.9	125.5
Gross Output	MW	272.8	277	277.7	280.5	282.9	303.5
Gross Efficiency (LHV)	%	57.7	58.6	58.7	59.3	59.8	57.9
Auxiliary Consumption	MW	4.1	4.5	4.5	4.6	5.2	5.0
Net Output	MW	268.7	272.5	273.2	275.9.	277.7	298.5
Net Efficiency (LIIV)	%	56.8	57.6	57.8	58.3	58.7	57.0
Net Heat Rate (LHV)	kJ/kWh	6,337	6,249	6,233	6,172	6,132	6,320
Net Heat Rate (LHV)	Btu/kWh	6,006	5,923	5,908	5,850	5,812	5,990

The above examples illustrate how the net efficiency can rise from 56.8% to 58.7% and the output from 268.7 MW to 277.7 MW by changing the concept of the water/steam cycle. Adding supplementary firing can increase this output further, but for cycles with several pressure levels at the expense of the efficiency.

Selection of a Combined-Cycle Concept

This section examines what factors must be considered when choosing a combined cycle concept for a future power plant, Figure (4-50). There are three main parts to the process:

- Defining the requirements
- Assessing the influence of site conditions
- Determining the solution

Defining the Requirements

The requirements define what a plant should be able to achieve in terms of performance and operational capability. The most significant requirement is the power demand. When this is identified, it is important to determine any limits above or below the nominal power rating, which may be imposed, for example, by grid limitations or the need to meet an internal auxiliary power demand. The best "fit" of turbines corresponding to this range must then be found taking into consideration the base load point at which the gas turbines will be operating.

If there is a process demand it will be necessary to look at the temperature, pressure, and mass flow that will be required at the supply limit. The allowable fluctuation in these conditions and the demand variation over time should be known, in order to design the plant for all possible operating conditions. If there is a return of steam or condensate from the process its condition and quality need to be implemented into the cycle design as well.

Combined-cycles are intended for various uses, such as baseload, cycling, or frequency-following. In order to account for such factors at the design stage it is important to have information about expected operating hours and number of starts per year.

As independent power producers (IPPs) come into the market, financing becomes more and more important. When utilities were the main purchasers of power plants, they could often finance power stations themselves off their balance-sheet. Today, in order to be worthy of credit, any new power plant must be able to compete with other power plants on the open market on the basis of $/MWh hour generated—not just for the first year but throughout the pay-back period of the plant and preferably even longer.

Analysing the Site-Related Factors

Site-related factors are specific to the intended location of the combined-cycle plant and are usually out of the control of the power plant purchaser. Those that affect the choice of the cycle must be considered in the cycle-selection process. One fundamental factor, which usually is country dependent, is whether the local electricity grid is rated at 50 or 60 Hz. This affects the selection of the gas-turbine type

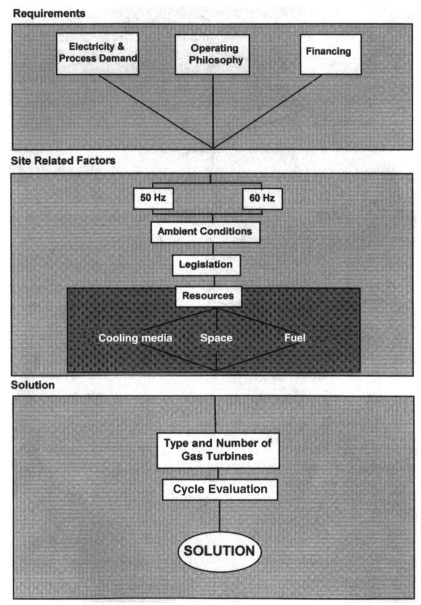

Figure 4-50 Selection of a Combined-Cycle Concept

because larger gas turbines are designed for specific frequencies. Smaller ones are usually geared to operate with both of these frequencies.

Ambient Conditions

Some ambient conditions affect the performance of the cycle. This section looks at the influences of these conditions as the design point of the cycle changes. How an already dimensioned combined-cycle plant behaves with different ambient conditions will be discussed in chapter 7.

The gas turbine is a standardized machine, so that a given machine is used for widely different ambient conditions. This can be justified economically because a gas turbine that has been optimized for an air temperature of 15°C (59°F) does not look significantly different from one that has been designed for, say, 40°C (104°F). The costs for developing a new machine could not be justified. Manufacturers quote gas-turbine performances at ISO ambient conditions of 15°C, (59°F), 1.013 bar, (14.7 psia) and 60% relative humidity. The gas turbine will perform differently at different ambient conditions and this will have an effect on the steam process.

Unlike the gas turbine, the steam turbine is usually designed for a specific application. The exhaust steam section design, for example, depends on the condenser pressure at the design point, e.g. the exhaust section that would be chosen for a condenser pressure of, say, 0.2 bar (5.9"Hg) can no longer function optimally if the pressure is only 0.045 bar (1.3"Hg). Also, blade path design in a steam turbine depends on the live-steam pressure, which is not the same for all cycles.

The main ambient conditions important here are the air temperature and pressure. Relative humidity also has a minor influence but becomes more important if the water for cooling the condenser is re-cooled in a wet cooling tower.

Ambient air temperature

There are three reasons why the air temperature has a significant influence on the power output and efficiency of the gas turbine.

- Gas turbines always draw in a constant volume flow to the compressor. Increasing the ambient air temperature reduces the density of the air and thereby reduces the air mass flow contained in the given volume flow. The air mass flow determines the gas turbine output at a given turbine inlet temperature, TIT, and pressure ratio

- The specific volume of the air increases in proportion to the intake temperature (in K), increasing the power consumed by the compressor, without a corresponding increase in the output from the turbine

- As the air temperature rises and the mass flow decreases the pressure ratio within the gas turbine is reduced. This is because, as the swallowing capacity of the gas turbine is given, the law of cones reduces the pressure before the turbine. The same principle applies inversely to the compressor, but because the turbine is dominant, the total balance is negative

Figure 4-51 shows the gas turbine characteristic at two different ambient temperatures in a temperature/entropy diagram. The exhaust gas temperature is higher as the air temperature increases because the turbine pressure ratio is reduced while the TIT remains constant. The result is a decrease in gas turbine efficiency and output as ambient temperature rises. However, the effect on the performance of the combined-cycle as a whole is more moderate because a higher exhaust gas temperature improves the performance of the steam cycle.

Figure 4-52 shows the relative efficiencies of the gas turbine, steam process, and the combined-cycle plant as a function of the air temperature, with other ambient conditions and the condenser pressure remaining unchanged. In the diagram an increase in air temperature has a slightly positive effect on the efficiency of the combined-cycle plant, since the increased temperature in the gas turbine exhaust raises the efficiency of the steam process enough to more than compensate for the reduced efficiency of the gas turbine unit. This behavior is not surprising considering the Carnot efficiency (Eq. 3-1). The increase in the compressor outlet temperature causes a slight increase in the average temperature of the heat supplied (T_E), as well. Most of the exhaust heat is dissipated in the condenser so the cold temperature (T_A) does not change because the condenser vacuum is constant. The overall efficiency of the combined-cycle plant will increase.

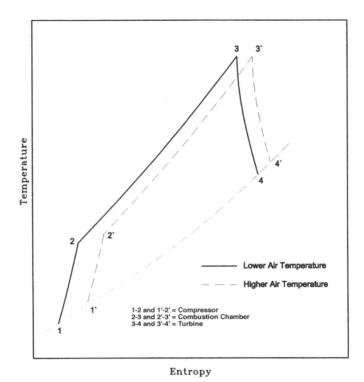

Entropy

Figure 4-51 Entropy/Temperature Diagram for a Gas Turbine Process at Two Different Ambient Air Temperatures

Figure 4-53 shows how the power output of the gas turbine, steam turbine, and combined-cycle decrease with an increase in the air temperature. The effect is less pronounced for the combined-cycle than for the gas turbine alone. The effect on power output of the combined-cycle is more marked than that on the efficiency because changes in the mass flows of air and exhaust gases are more dominant than changes in the exhaust gas temperature.

Ambient air pressure

Gas turbine performance is normally quoted in the literature at an air pressure of 1.013 bar (14.7 psia). This corresponds approximately to the average pressure prevailing at sea level. A different site elevation and daily weather variations result in a different average air

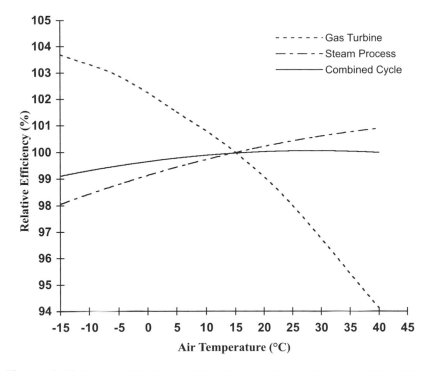

Figure 4-52 Relative Efficiency of Gas Turbine, Steam Process and Combined-Cycle as a Function of the Air Temperature at constant vacuum

pressure. Figure 4-54 shows the relationship between site elevation and ambient air pressure, and how this influences the relative output of the gas turbine, steam turbine and the combined-cycle. The air pressure has no effect on the efficiency at a given ambient temperature even though the ambient air pressure has an influence on the air density similar to that of the air temperature. At a lower ambient pressure the back pressure of the gas turbine is correspondingly lower–not considering inlet and outlet pressure drops. This leaves no influence on the gas turbine process except that of the reduced mass flow. Assuming that no change takes place in the efficiency of the steam process (which corresponds quite well to the real situation), there is the same variation in the power output of the steam turbine and hence the combined cycle.

Because the power outputs of the gas turbine and the steam turbine vary in proportion to the air pressure, total power output of the

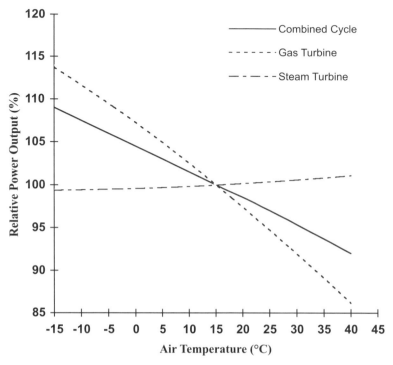

Figure 4-53 Relative Power Output of Gas Turbine, Steam Turbine and Combined-Cycle as a Function of Air Temperature at Constant Vacuum

combined-cycle plant varies in proportion. The fact that the gas turbine inlet and outlet pressure drops were held constant for the calculations in Figure 4-54 accounts for the slight difference in the relative power output compared to the air pressure. The efficiency of the plant remains constant, however, since both the thermal energy supplied and the air flow vary in proportion to the air pressure.

Ambient relative humidity

Figure 4-55 shows that gas turbine and combined-cycle output will increase if the relative humidity of the ambient air increases with other conditions remaining constant. This is because at higher levels of humidity, there will be a higher water content in the working medium of the gas cycle resulting in a better gas turbine enthalpy drop and more exhaust gas energy entering the HRSG.

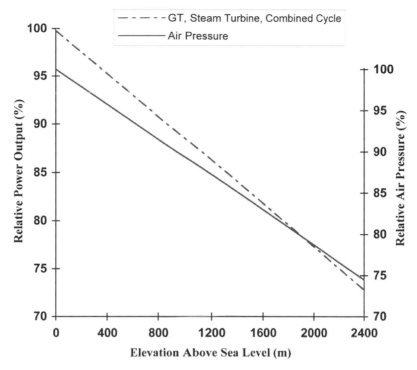

Figure 4-54 Relative Power Output of Gas Turbine, Steam Turbine and Combined-Cycle and Relative Air Pressure versus Elevation Above Sea Level

There is a further influence for plants with cooling towers, where relative humidity is directly related to the condenser vacuum and hence the steam turbine exhaust temperature. In these cases, a lower humidity results in a better vacuum. This is discussed in the section entitled "Resources", which follows.

Legislation

Many environmental considerations affect the design and building of power stations of all types. Of these, the choice of a combined-cycle concept is mainly influenced by legislation which limits emission levels (especially of nitrous oxides (NO_x)) released into the atmosphere. Current dry low-NO_x gas turbine combustors can often

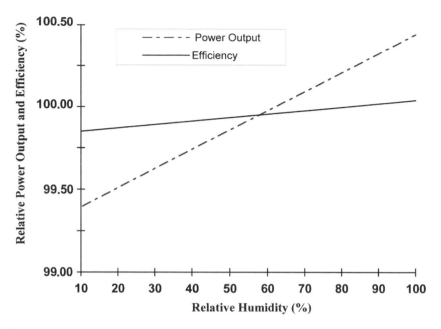

Figure 4-55 Relative Power Output and Efficiency of Gas Turbine and Combined-Cycle versus Relative Humidity at Constant Vacuum

achieve the required levels on gaseous fuel but special measures may have to be taken if oil fuels are fired.

One way to reduce the formation of NO_x during combustion is to lower the temperature of the flame, since the speed of the reaction producing NO_x is noticeably rapid only at very high temperatures. The common approach for plants burning gaseous fuels is to use premix burners to lower the flame temperature by mixing air at the outlet of the compressor with the fuel before it is ignited in a vortex-breakdown zone. Injecting water or steam into the combustor can produce the desired temperature reduction but it will affect the output and efficiency of the gas turbine. This is therefore mainly used for fuel oil, where premix is problematic due to the risk of pre-ignition. The amount of NO_x reduction possible depends on the water or steam-to-fuel ratio as described in chapter 9.

For gas turbines in simple-cycle with no HRSG (and, therefore, no available steam source), water is injected for the NO_x control but this has a negative effect on efficiency. The problem with steam injection is finding steam at a suitable pressure level. Generally HP live-steam pressure is too high and IP and LP steam pressures too low. For large industrial gas turbines the pressure level required for steam injection is between 30 and 50 bar (420 and 710 psig), depending on the gas turbine and the load. Using reduced HP steam is usually the simplest and the least expensive solution but is exergetically undesirable.

For this reason, a base load plant with gas-turbine steam injection will usually have a steam extraction in the steam turbine at the appropriate pressure level with a back-up from the HP live-steam line for those operating points at which the extraction pressure is too low. Steam extraction is, however, more problematic in combined-cycle blocks with several gas turbines. Unless all of the gas turbines are in operation, the pressure at the extraction point decreases to where it is in most cases inadequate. It is then necessary to switch over to the live-steam, which negatively affects the efficiency.

Power output is increased by both water and steam injection due to the resulting increase in the mass flow through the gas turbine. The water injection has the greater effect on output for the same injection mass flow because with steam injection, steam turbine output is decreased. Water or steam injection is also a means of power augmentation. Efficiency of a combined-cycle plant is decreased in both cases, however–less so by steam than by water, because steam brings more internal energy to the combustor. For the same reason if the water is hotter the detriment in efficiency is less.

Figure 4-56 shows the effect that injecting water or steam has on the output and efficiency of a dual-pressure cycle. The steam is extracted from the steam turbine, the water from the makeup water line, at 15°C (59°F) or at the exit from the economizer at 150°C (302°F). The amount of water or steam is provided by the ratio of water/steam to fuel, which is determined by the level of NO_x reduction necessary or, sometimes, by the amount of power augmentation required.

Apart from NO_x, other emissions are often restricted–CO, particulates, and so forth–but these influence the component design rather than the cycle performance and so do not influence the selection of the concept.

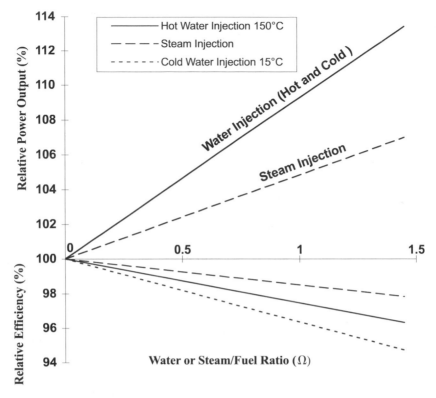

Figure 4-56 Effect of Water and Steam Injection on Relative Combined-Cycle Power Output and Efficiency versus the Water or Steam/Fuel Ratio

Resources

At every site the resources that are available are different. Cooling media and fuel are the main resources affecting the selection of the cycle concept. Although space is also a resource, it affects the cycle only when space is restricted by, for example, the need to use existing buildings or part of an existing cycle. This may lower the capital cost of the plant but could also limit the possibilities for performance improvements by, for example, restricting the size of the HRSG or influencing the choice of shaft configuration.

Cooling media

To condense the steam, a cooling medium must carry off waste heat from the condenser. Generally this is water, which has a high specific thermal capacity and good heat-transfer properties. Where water is available, cooling can be done in a direct system (using water drawn from a river or the sea) or in a wet-cooling tower. Where water is not available, very expensive, or to facilitate permitting an air-cooled condenser is used. These are expensive items that require a lot of auxiliary power and operate at a poorer vacuum than the water cooled options.

The temperature of the cooling medium has a major effect on the efficiency of the thermal process. The lower this temperature is, the higher the efficiency that can be attained because the pressure in the condenser is lower, producing a greater useful enthalpy drop in the steam turbine and hence an increase in steam turbine output. This is illustrated in Figure 4-57.

The trend is similar for single-, dual-, and triple-pressure cycles. This effect is much less significant above 100 mbar (3.0"Hg) as the relative change in the pressure decreases and there is less impact on the steam-turbine enthalpy drop. However, plant costs are reduced if the pressure is higher, due to the lower volume flow of exhaust-steam resulting in a smaller steam turbine.

Figure 4-58 shows typical condenser pressure values as a function of the temperature of the cooling medium for direct water-cooling, water-cooling with a wet-cooling tower, and direct air-cooling. The best vacuums are attained with direct-water cooling, the worst with direct condensation with air. In the comparison, it must be remembered that the water temperature is generally lower than the corresponding ambient air temperature. The curves for direct cooling and cooling tower cooling are calculated with cooling water temperature rises of 9 K, (16°R) and 13 K, (23°R) respectively.

For the wet-cooling tower the wet-bulb temperature is given as the temperature of the cooling medium. This is a function of the dry-bulb temperature and the relative humidity and can be read from an enthalpy-moisture content diagram for air.

Using Figure 4-57 in combination with Figure 4-58, the influence on the steam turbine output of the various cooling media can be seen.

Fuel

It has already been shown how the sulfur content in the fuel influences the cycle concept because of the feedwater temperature. The fuel type and composition also have a direct influence on gas-turbine performance and the emissions produced. Power plants often have a main fuel and a back up fuel.

The lower heating value (LHV) of the fuel is important because it defines the mass flow of fuel that must be supplied to the gas turbine. The lower the LHV the higher the mass flow of fuel required to provide a certain chemical heat input, normally resulting in a higher power output and efficiency. This is why so-called low BTU gases can result in high power outputs.

The fuel composition is equally important in influencing the performance of the cycle because it will determine the enthalpy of the gas entering the gas turbine and hence the available enthalpy drop and gas-turbine output. This is the reason why the influence of the fuel on performance cannot be given as a function of LHV only.

One way of improving the efficiency of the cycle is to raise the apparent heating valve (LHV + sensible heat) of the fuel by preheating it with hot water from the economizer of the HRSG. This also improves heat utilization in the HRSG because additional water is heated in the economizer. The fuel consumption is correspondingly reduced, because each unit mass of fuel contains more sensible heat, so a lower fuel mass flow is required. Figure 4-59 shows schematically how hot water is extracted from the economizer (5) of a dual-pressure HRSG. The cooled water (4) returns to the feedwater tank. Preheating natural gas fuel from 15°C (59°F) to 150°C (302°F) can improve the cycle efficiency by approximately 0.7% (relative).

Another factor that may be important for gas fired plants is the available gas supply pressure. The gas turbine requires a certain pressure depending on the design of the burners in the combustor and the

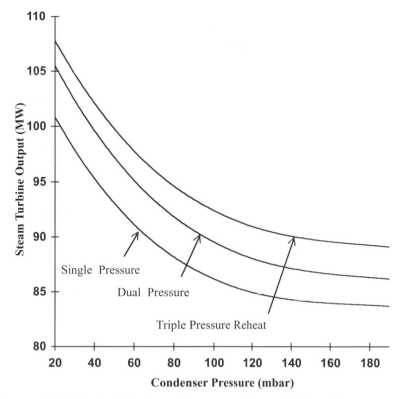

Figure 4-57 Effect of Condenser Pressure on Steam Turbine Output

gas-turbine pressure ratio. Sometimes a gas compressor is required to increase the supply pressure and this will increase the temperature of the fuel in proportion to the compression ratio. The benefit of a pre-heater in such cases is reduced perhaps rendering it economically un-viable, because the efficiency improvement is too slight to justify the additional investment in the water/gas heat exchanger, HRSG surface and piping.

The following table lists fuels that can be fired in a gas turbine, all of which are liquid or gas fuels:

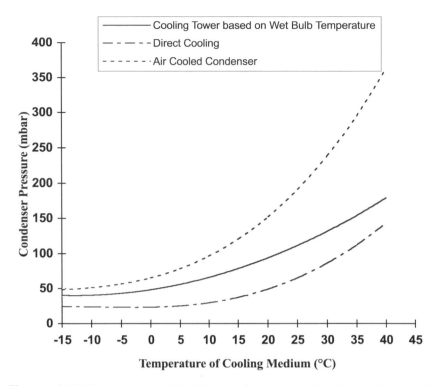

Figure 4-58 Temperature of Cooling Medium versus Condenser Pressure for Direct Cooling, Wet Cooling Tower and Air Cooled Condenser

Table 4--2 Possible Fuels for Combined-Cycle Applications

a) Standard Fuels	b) Special liquid fuels	c) Special gas fuels
Natural gas	Methanol / Kerosene Naphtha	Synthetic gas
Diesel oil	Crude oil	Blast furnace gas
	Heavy oil, residuals	Coal gas with medium or low heating value
	Oil shale oil	

Note: The use of categories (b) and (c) is limited, depending upon the exact chemical analysis and the type of gas turbine involved. Generally, industrial gas turbines with large combustors are better able to handle these fuels than modern gas turbines with high turbine inlet temperatures.

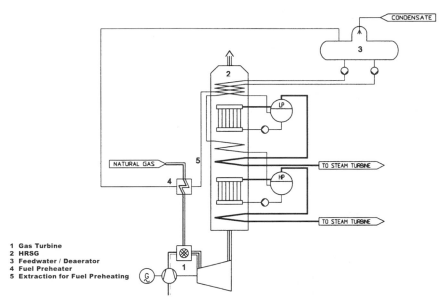

Figure 4-59 Flow Diagram to Show Fuel Preheating

Determining the Solution

When plant requirements and site data are known, a concept for the cycle can be determined. In doing this, both technical and economic aspects must be taken into account.

In recent years, globalization of both the electricity generation industry and fuel prices, along with financing possibilities, has led to markets that are undifferentiated, with similar requirements coming from customers in completely different parts of the world. This has enabled manufacturers to develop standard cycles for the applications that can be adapted to meet the specific needs of a given project.

Selection of the Gas Turbine

The first stage of cycle selection is determining the size and number of gas turbines to use based on the need to meet a certain power demand and process-steam requirement, if there is one.

For a plant with a given gas turbine, supplying process steam will decrease power output because steam is removed from the cycle rather than being used for power generation. This may mean choosing a larger gas turbine than would be the case if the process were not required. A gas turbine of a lower rating than required could still be chosen for a given application with power shortfalls covered by additional steam turbine output through supplementary firing. It makes no difference to the steam turbine whether exhaust gas temperatures are attained directly from the gas turbine or by means of supplementary firing. However, from the point of view of efficiency, it is better to convert the energy directly in the gas turbine at a higher exergetic level than in the HRSG duct burner. An alternative means of power augmentation is water or steam injection into the gas turbine as described earlier.

If more than one gas turbine is needed, a choice must be made between a multi-shaft configuration with all of the gas turbines and HRSGs feeding the same turbine or several single-shaft blocks each of which has a steam turbine and gas turbine sharing the same generator. The latter provides flexibility at part-load operation. Several manufacturers have developed standard single-shaft power plants for their gas turbines. These standardized units provide advantages related to fast installation times and availability.

Not all gas turbines with the same inlet air flow have the same performance. It depends on the turbine inlet temperature (TIT), the design concept (i.e., one or two stages of combustion), and resulting exhaust gas conditions. Since this has a marked influence on the performance of the combined-cycle, it is worth looking at in more detail.

When the gas turbine exhaust temperature is lowered, both the thermodynamic quality of the steam process (seen by the steam turbine output) and the HRSG efficiency decrease, as shown in Figure 4-60. The effect is more pronounced with the single-pressure than with a dual-pressure cycle because the energy utilization rate falls off more quickly.

Figure 4-61 shows the ratio between the output of the dual-pressure and single-pressure cycles as a function of the gas-turbine exhaust-gas temperatures. The lower the gas turbine exhaust gas temperature, the more sense a dual-pressure system makes. At a theoretical exhaust gas temperature of 750°C (1382°F), this ratio is practically equal to 1. A similar influence is seen for the triple-pressure cycle.

The dual-pressure process therefore provides no advantage over the single- pressure process for exhaust-gas temperatures above 750°C (1382°F). This fact is important in cycles with supplementary firing.

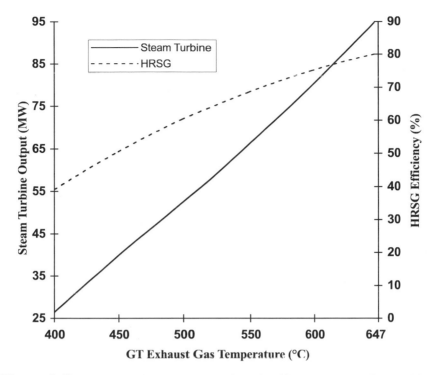

Figure 4-60 Steam Turbine Output and HRSG Efficiency versus Gas Turbine Exhaust Temperature for a Single Pressure Cycle

Cycle Evaluation

When defining a cycle concept for a power plant there is always a trade-off between the performance, the required investment, and the economic criteria. The performance and the investment are directly related to each other because as the complexity of a cycle increases there is generally an increase in performance but also in the required investment. The exact relationship depends on the type of gas turbine–or rather, the exhaust gas conditions of the gas turbine entering the water steam cycle.

Figures 4-62 and 4-63 show the relative performance for different concepts plotted against the relative turnkey price for cycles based on two different 178 MW class gas turbines, with exhaust gas temperatures of 647°C, (1197°F) and 525°C, (977°F) respectively.

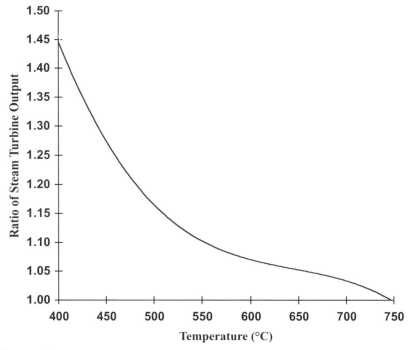

Figure 4-61 Ratio of Steam Turbine Output of a Dual Pressure to that of a Single Pressure Cycle as a Function of Gas Turbine Exhaust Temperature

Though the trends of the curves are similar, their scales are quite different. The addition of more pressure levels to the gas turbine with high exhaust-gas temperature can improve the efficiency by almost 3.5% whereas the improvement for the gas turbine with the low exhaust gas temperature is almost 7%. This is because with the second gas turbine, there is more low grade heat available for additional pressure levels in the HRSG due to the exhaust gas temperature, as was explained earlier in this chapter. However, the improvement from the dual-pressure cycle to the most efficient cycle is almost the same in both cases. The shift in optimum efficiency between the dual-pressure reheat and triple-pressure reheat cycles for the two gas turbines is also interesting.

For the gas turbine with the lower exhaust gas temperature in the triple-pressure reheat cycle, the IP pressure part still makes a sig-

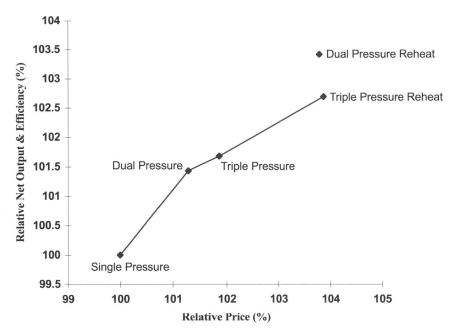

Figure 4-62 Indicative Relative Price versus Performance of Different Combined-Cycles based on a 178 MW Class Gas Turbine with an Exhaust Gas Temperature of 647°C (1197°F)

nificant contribution to the steam turbine output, although the HP contribution is dominant.

With the other gas turbine the HP contribution is even greater. The dual-pressure reheat cycle uses a higher live-steam pressure which more than compensates for the lost IP production.

Any of the concepts could be appropriate for a power plant and the final choice depends on the economic criteria for the specific project conditions. These are mainly the criteria related to the cost of electricity described in chapter 2. The example using equation 2-8 shows that for an assumed set of economic criteria, it is worthwhile investing up to 3.3% more capital to achieve a 1% improvement in performance. Applying this for the gas turbine in the examples, it can be seen from Figure 4-62 that this leads to a dual-pressure reheat cycle.

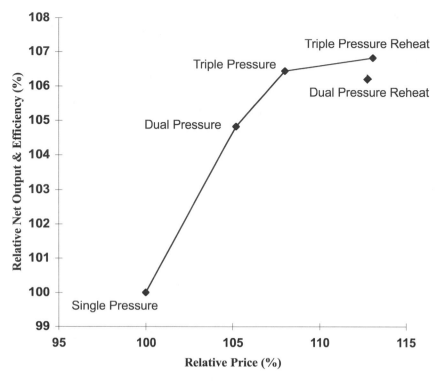

Figure 4-63 Indicative Relative Price versus Performance of Different Combined-Cycles based on a 178 MW Class Gas Turbine with an Exhaust Gas Temperature of 525°C (977°F)

Conclusion

Using this approach, then, the final solution of Figure 4-50 can be reached. With other economic criteria or with a different gas turbine the solution is different due to the many factors influencing the selection process. The tendency towards globalization of the requirements means that these factors are tending to converge. The result is the emergence of concepts as ideal solutions for certain sets of conditions and a movement towards standardization of these concepts. This brings with it advantages for the investor, such as easier permitting, shorter installation times, and lower risks due to proven equipment. The result is that these standard cycles can cover a wider variation of economic criteria, enhancing further the benefits of the standardization process.

APPLICATIONS OF COMBINED-CYCLES

CHAPTER 5

5

Applications of Combined-Cycles

The combined-cycles described in chapter 4 are purely for electricity generation. In this chapter we explore how the thermodynamic advantages of combining gas and steam cycles can apply for other purposes–the production of thermal energy or the repowering of existing steam power plants by integrating a gas cycle into the existing steam cycle. Some other, less conventional concepts are also discussed, though most of these have as yet limited commercial usage.

Cogeneration

Cogeneration means the simultaneous production of electrical and thermal energy in the same power plant. The thermal energy is usually in the form of steam or hot water. The types of cogeneration plant discussed in this chapter fall into three main categories:

- industrial power stations supplying a process requirement
- district heating power plants
- power plants coupled to seawater desalination plants

The thermodynamic superiority of a combined-cycle plant over a conventional power plant is even more pronounced in cogeneration plants than in plants used to generate electricity alone. The difference between the average temperature of heat added and heat dissipated is greater in a combined-cycle than in a steam cycle (see chapter 3). If both types of power plant now supply heat at the same temperature level, the reduction in this temperature difference is the same for both, but the relative reduction in the combined-cycle is smaller because here the total temperature difference is larger.

A cogeneration plant may also have supplementary firing. This offers much greater design and operating flexibility because the steam production can be controlled, independently of the electrical power output, by regulating the fuel input to the supplementary burner. The power output is controlled by the gas turbine. The cycle efficiency will, however, normally be lower if supplementary firing is used, not always making this a desirable solution for plants where the owner must compete to sell electricity to the grid.

Thermal energy, in the form of steam can be extracted from a point in the HRSG, in the live steam piping or from the steam turbine. The optimum extraction point depends on the required steam pressure, temperature and quantity over the load range and is very important in defining the type of cycle which should be used. In the following sections the advantages and disadvantages of the various possibilities are discussed.

Industrial power stations

Wherever both electrical power and process steam are needed, it is thermodynamically and usually also economically better to produce both products in a single plant. The number of possible concepts is large because each plant is to a certain extent unique. The following examples have only one process steam supply. Often several processes are required, at different conditions, but the basic considerations remain unchanged. There are three basic extraction concepts that relate to the steam turbine:

- cycle with back-pressure steam turbine
- cycle with extraction/condensing steam turbine
- cycle with no steam turbine

A simplified flow diagram for a cycle with a back-pressure steam turbine is shown in Figure 5-1. The steam turbine (7) is designed so that the exhaust pressure matches the process requirement (3). A by-pass (5) around the steam turbine is designed to reduce the live steam to the required pressure and temperature level for the process in case the steam turbine is out of operation and the process must still be supplied. In this cycle all of the steam generated in the HRSG goes into the process, except for a small quantity used for feedwater tank heating and deaeration.

1 Gas Turbine
2 HRSG
3 Process Supply
4 Feedwater / Tank
  Deaerator
5 Pressure Reducing Station
6 Supplementary Firing
7 Back Pressure Steam Turbine
8 Pegging Steam Line
9 Feedwater Pump

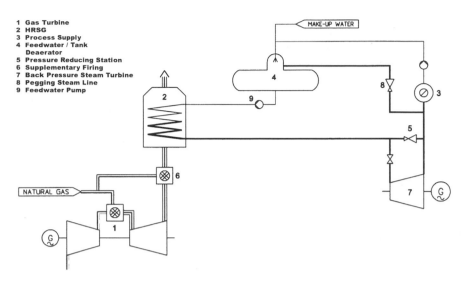

Figure 5-1 Simplified Flow Diagram of a Cogeneration Cycle with a Back Pressure Turbine

For such a cycle all or some of the steam lost to the process may be returned as condensate. Any loss in mass flow must be replaced with make-up water that usually enters the cycle at the deaerator because it is fully oxygenated. If condensate is returned from an industrial process the quality may be poor and it may have to be treated before re-entering the water/steam cycle to achieve a suitable quality. If the return condensate is at a high temperature this may influence the temperature at which the feedwater tank operates or the amount of heating steam to be extracted from the steam turbine in order to get sufficient deaeration and feedwater heating.

Figure 5-2 shows a concept using an extraction/condensing turbine (6). Here, the process steam is extracted from the steam turbine. The process (5), which must be held at a constant pressure level, is regulated internally (4) within the steam turbine. A back-up supply for the process is provided through the live steam reduction station (7). The steam, which is not extracted for the process further, expands in the turbine and is condensed in a cold condenser (8), as for a non-cogeneration plant. For such cycles even without supplementary firing the process steam flow can be varied without affecting the gas turbine load. This provides some additional operating flexibility, but not to the

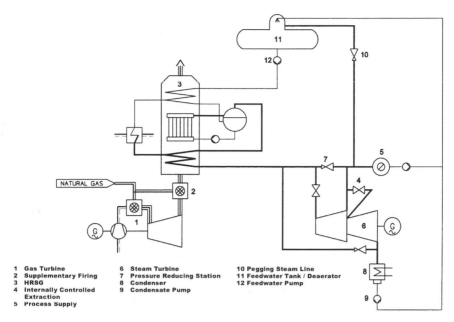

Figure 5-2 Flow Diagram of a Cogeneration Cycle with an Extraction/Condensing Steam Turbine

1 Gas Turbine	6 Steam Turbine	10 Pegging Steam Line
2 Supplementary Firing	7 Pressure Reducing Station	11 Feedwater Tank / Deaerator
3 HRSG	8 Condenser	12 Feedwater Pump
4 Internally Controlled	9 Condensate Pump	
Extraction		
5 Process Supply		

same extent as supplementary firing. This is because for the same gas turbine output, the steam turbine output and the process extraction flow are interdependent in a smaller range.

A cycle with no steam turbine is shown in Figure 5-3. This is appropriate if a large quantity of process steam is required at a high pressure and temperature, leaving little possibility for steam turbine expansion. There is supplementary firing in order to achieve sufficient process steam mass flow. The steam goes directly from the HRSG to the process (5) and the HRSG is designed to generate steam at the process conditions. The pressure may be controlled by a regulating station (4) but it is quite common for the main pressure control to take place within the plant to which the process steam is being exported.

Evaluation of a cogeneration cycle

The performance of a cogeneration cycle is defined not only by the efficiency but also by parameters such as fuel utilization and power

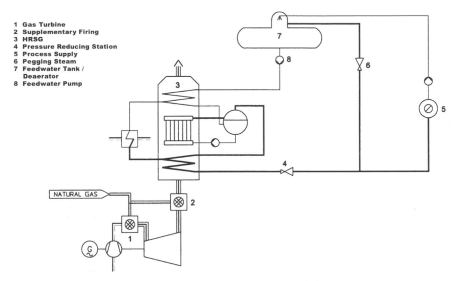

1 Gas Turbine
2 Supplementary Firing
3 HRSG
4 Pressure Reducing Station
5 Process Supply
6 Pegging Steam
7 Feedwater Tank /
   Deaerator
8 Feedwater Pump

NATURAL GAS

Figure 5-3 Flow Diagram of a Cogeneration Cycle with no Steam Turbine

coefficient. These parameters take into account the thermal as well as the electrical output.

Fuel utilization is a measure of how much of the fuel energy supplied is usefully used in the plant. It is equal to the sum of electrical output and thermal output divided by the fuel input.

The power coefficient, (also called the alpha-value) is defined as the ratio between the electrical power and the thermal energy produced. Combined-cycles tend to have high power coefficients so they are more likely to be used for cogeneration applications with a relatively high power demand. This is because the electrical output of the gas turbine—about two thirds of the total plant output—cannot be converted into thermal energy. In a conventional steam power plant all of the energy produced could be exported to the process giving a possible power coefficient of zero.

Figure 5-4 is a heat balance for a cycle with a back pressure turbine using the 178 MW gas turbine and a single-pressure cycle with supplementary firing to 750°C, (1,382°F). The exhaust pressure of the steam turbine (and hence the process pressure) is 3 bar (29 psig). Condensate returns from the process at 50°C (122°F) and this assists with the feedwater preheating so that only a small amount of steam is taken from the steam turbine exhaust for feedwater heating to achieve a feed-water temperature of 60°C (140°F).

The cycle is directly comparable to Figure 4-49 (single-pressure cycle with supplementary firing for electricity generation only), except that there is a cold condenser in Figure 4-49 and in Figure 5-4, heating steam is taken from the process. Performance of the HRSG is the same in both cycles with identical stack temperatures and steam flows. Due to the high turbine back pressure, the output in the cogeneration cycle is 50.8 MW less with an electrical efficiency of 48.3% compared to the full condensing cycle's 58%. However, 239.7 MJ/s of thermal energy is exported to the process. This results in a fuel utilization of 93.9%. The remaining 6.1% of the energy is lost through the stack (5.5%), in radiation losses, and in generator and mechanical losses in the steam turbine and gas turbine. There is no condenser loss because all of the steam is used in the process. The power coefficient is 105%.

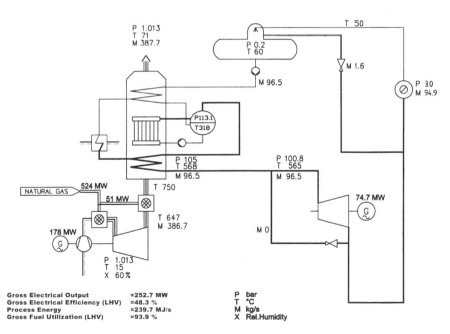

Gross Electrical Output	=252.7 MW
Gross Electrical Efficiency (LHV)	=48.3 %
Process Energy	=239.7 MJ/s
Gross Fuel Utilization (LHV)	=93.9 %

P	bar
T	°C
M	kg/s
X	Rel.Humidity

Figure 5-4 Heat Balance for a Single Pressure Cogeneration Cycle with Supplementary Firing

The power coefficient of a plant is largely affected by three pa-rameters:

- amount of fuel supplied directly to the boiler
- size of the condensing portion of an extraction/condensing turbine
- pressure level of the process steam

Figure 5-5 shows the effect of process steam pressure on the rel-ative power output and power coefficient for a combined-cycle with a back-pressure steam turbine. The reference point at 100% relative power output is equivalent to a condensing cycle. The higher the process steam pressure, the less electricity is produced because there is a smaller enthalpy drop in the steam turbine. In theory, the power co-efficient approaches 170% for back-pressures equal to those of con-densing applications. As the pressure increases the power coefficient

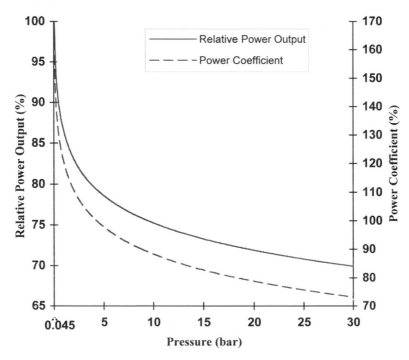

Figure 5-5 Effect of Process Steam Pressure on Relative Combined-Cycle Power Output and Power Coefficient for a Single Pressure Cycle with 750°C Sup-plementary Firing

falls exponentially to the limit set by the gas turbine output because the steam turbine electrical output is falling and the thermal output rising at the same time.

As the power coefficient increases, the electrical efficiency approaches asymptotically that of a fully condensing combined-cycle plant (Fig. 5–6). The fuel utilization remains constant because in a plant with no condenser almost all of the steam entering the steam turbine is usefully used and the energy is just shifted between process energy and electrical output by varying the steam turbine back pressure.

Where supplementary firing is installed, additional steam generation contributes to the electrical output of the steam turbine but has no corresponding gas turbine output. Thermal output therefore increases at a much greater rate than electrical output with increasing supplementary firing temperature resulting in a decrease in the power coefficient.

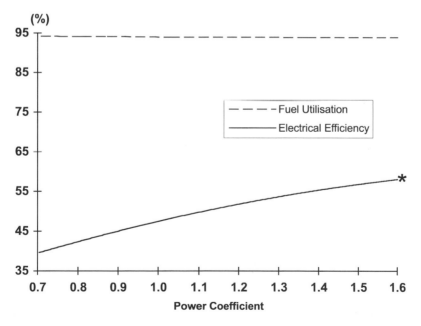

* **Limit for Cogeneration (=Condensing) Application**

Figure 5-6 Effect of Power Coefficient on Electrical Efficiency and Fuel Utilization for a Single Pressure Cycle with 750°C Supplementary Firing and Decreasing Steam Turbine Back Pressure

If higher power coefficients are required, an extraction/condensing steam turbine offers greater design and operating flexibility. The condensing portion of the turbine enables the electrical power output to be increased but only at the cost of thermal output and fuel utilization. If designed properly, such a cycle can operate at any point in the range from full condensing to full back-pressure mode. For back-pressure mode, cooling steam may be required for the condensing part of the steam turbine.

This solution is unrealistic for large plants where power coefficients must be varied. Where lower power coefficients are required, mixed process steam production using some steam from a back-pressure-turbine and the rest from an HRSG could be used though the efficiency of power production is lower.

The highest fuel utilization can be obtained with a cycle with a back-pressure turbine but the flexibility is poor because large quantities of steam are always generated unless the load of the gas turbine is reduced.

Main Design Parameters

Cogeneration plants exhibit the characteristics of non-cogeneration plants as described in chapter 4. In addition to these, however, the following must also be taken into account when designing for a steam-process application.

The live steam pressure for a plant with a steam turbine must be high enough to ensure a reasonably high enthalpy drop between the live steam and the process steam in order to optimize the steam turbine output. This is especially true if a relatively high-pressure level is required for the process steam. For higher live steam pressure levels, heat utilization is poor in the HRSG. This can be avoided by introducing a lower pressure level or supplementary firing to the cycle.

Figure 5-7 shows a cycle using an unfired dual pressure HRSG (2) and a back-pressure turbine (3). The LP steam is fed directly into the process steam system (4). The process supply is backed up with a reduction station (6) from the HP live-steam line. It may be necessary to regulate the temperature of the process by injecting feedwater as shown. For cycles without supplementary firing, additional pressure

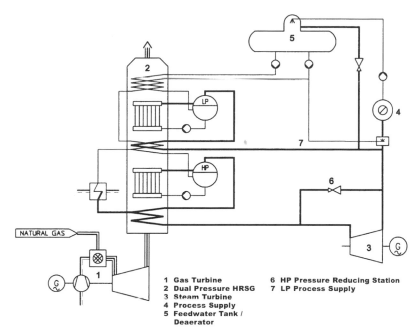

1 Gas Turbine   6 HP Pressure Reducing Station
2 Dual Pressure HRSG   7 LP Process Supply
3 Steam Turbine
4 Process Supply
5 Feedwater Tank /
  Deaerator

Figure 5-7 Flow Diagram of a Cogeneration Cycle with a Dual Pressure HRSG

levels improve the HRSG exhaust gas utilization, improving the overall cycle performance as explained in chapter 4. If oil were the fuel, the LP economizer would be omitted because a higher feedwater temperature would be required for this fuel.

District Heating Power Plants

A district heating system uses hot water to bring heat to towns and communities rather than using electrical power. A combined-cycle can be used as the heat source in the district heating network.

District heating water is required at temperature levels much lower than those used for processes in industrial power plants. Usual forwarding temperatures are between 80 and 150°C (176 and 302°F) with return temperatures of 40 to 70°C (104 to 158°F).

Steam extracted from the LP part of the steam turbine is used to heat this water so, for exergetic reasons, it is better in larger installations to divide the heating into two or even three stages. Figure 5-8 shows the temperatures of hot water and steam required in a one-stage

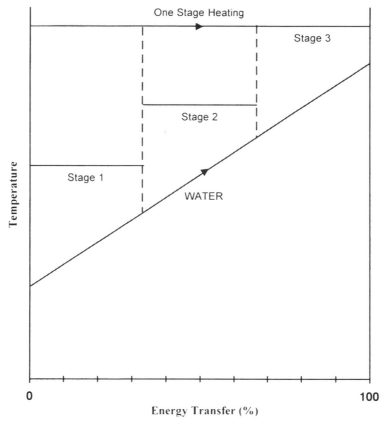

Figure 5-8 Comparison of 1-Stage and 3-Stage Heating of District Heating Water

and a three-stage process. Where a single stage is used, all steam must be extracted at the higher pressure level. This means there is less steam expanding in the steam turbine so the steam turbine power output is lower than that with a three-stage system. If only a small temperature increase in the district heating water is required the trend is towards a single stage of heating.

The main design criteria for district heating power plants are, therefore, the heat output and the water temperatures, which depend on ambient temperatures because these determine the demand for heating. The strong positive influence that a low air temperature has on power output is an advantage in this case, as maximum heating output is demanded when ambient temperatures are lowest. The design temperatures of the district heating water represent a compromise between maximum electrical output and low costs for transportation of the heat.

The cycle must have a high degree of operating flexibility, but must not become too complicated or too expensive. In particular, district heating power plants should not be designed for extreme conditions because this would make them too expensive and bring into question the economical feasibility.

In district heating power plants, independent control of electricity and heat production is generally not required. These power plants are usually integrated into large grids where other power stations are available for adjusting the total electrical power output to meet demand. Usually the only output parameter that must be regulated in the district heating plant is the thermal output.

Figure 5-9 shows the flow diagram and heat balance for a typical example using an extraction/back-pressure turbine with two stages of district heating in a dual pressure cycle (with the same HRSG as Fig. 4-22). The first stage of district heating is done using the steam turbine exhaust steam. The exhaust steam, at a pressure of 0.38 bar (11.2" Hg),

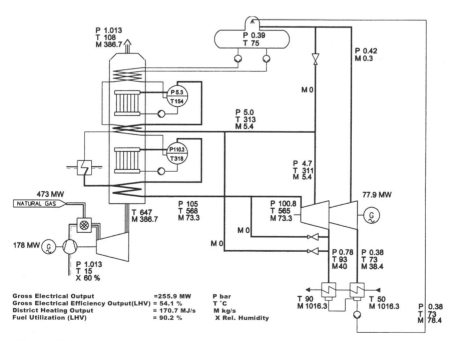

Figure 5-9 Heat Balance for a Cycle with Two Stages of District Heating

goes into the first district heating condenser where it is used to raise the temperature of the district heating water from 50 to 70°C (122 to 158°F). An extraction from the steam turbine at 0.78 bar (23" Hg) is used for the second stage of district heating, bringing water temperature up to 90°C (194°F).

Approximately half of the steam mass flow is used in each district heating condenser and at the design point it is optimal to have the same heat load in each stage, as is here the case. Condensate from the second-stage condenser cascades into the first stage condenser and is returned to the feedwater tank.

Compared to the condensing cycle of Figure 4-22, steam turbine output has fallen by 21.1 MW due to the district heating extractions. The electrical efficiency falls from 58.6% to 54.1%. The fuel utilization is 90.2% (lower than the example with a back-pressure turbine, mainly because of the increased feedwater temperature and resulting stack temperature).

Figure 5-10 shows a cycle using an extraction/condensing steam turbine. This time, each stage of district heating uses a steam turbine extraction and there is a cold condenser at the exhaust of the steam turbine into which the steam not used for district heating is further expanded. This type of cycle could be used if there was a need to increase the flexibility of operation between district heating and electricity generation.

Cycles Coupled with Seawater Desalination Units

In desalination plants, pure water is extracted from salt water (usually seawater). There are several possible methods but because the process involves distillation, the seawater must first be evaporated. A combined-cycle plant is a good source of heat for this evaporation process.

Combining a power station with a seawater desalination unit is particularly suited to combined-cycle plants because such power plants are generally built in oil- and natural gas-rich countries where ideal fuels for combined-cycles are readily available at a reasonable cost.

Larger desalination plants are usually designed to use the multistage "multi-flash" process. Seawater is heated in a row of cells with

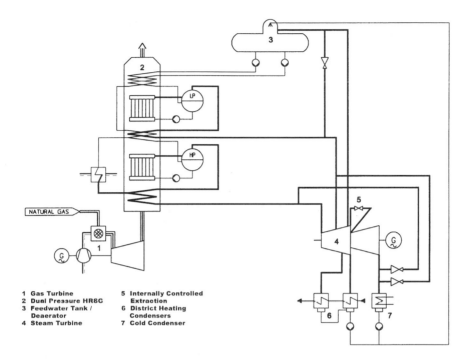

Figure 5-10 Flow Diagram of a District Heating/Condensing Cycle

the final stage of heating (at around 100°C (212°F)) taking place in a heater supplied with steam from the combined-cycle. Heated water returning through the cells heats incoming seawater in a counter-flow heat-exchange process. On entry to each cell the heated water undergoes a sudden pressure drop, causing instant evaporation of the water–known as "flashing". A part of the steam produced is then distilled and collected while the remainder recondenses, heating the cold water entering the cell before it moves on to the next stage. The heating steam is usually returned to the water/steam cycle in the form of condensate that–provided the quality is in order–can be admitted directly into the feedwater tank.

The maximum temperature of the water being heated is limited to prevent excessive formation of calcium carbonate ($CaCO_3$) deposits. This limit is between 90°C (194°F) and 120°C (248°F). Corresponding heating-steam pressures are between 1 and 2.5 bar (0 and 22 psig), which are ideal for a combined-cycle plant because a back-pressure steam turbine with a high enthalpy drop can be used, ensuring a high electrical output.

The electrical power output and the flow of process steam must usually be controlled independently of one another, so it is appropriate to install supplementary firing or to link the combined-cycle to a steam turbine condensing plant. Figure 5-11 shows a typical flow diagram based on a single pressure, supplementary fired HRSG and a "multi-flash" desalination system.

Heating steam is directly exhausted from a back-pressure turbine (2) to the desalination unit (5) where it is used to heat the seawater. The steam condenses in the process and is returned to the feed water tank (6). If the steam turbine is out of operation, supply for the desalination steam is provided through a pressure-reduction station (7). The high quantities of export steam at a high specific volume result in very large diameter pipelines to the desalination unit. For this reason, the steam may be exported in more than one parallel stream that is more cost-effective due to smaller piping and valving.

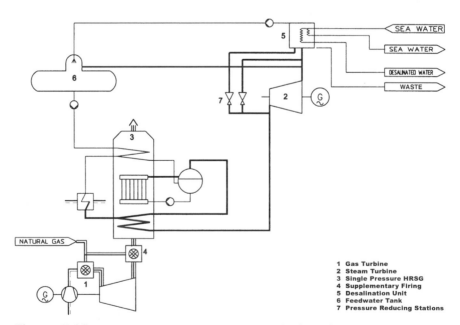

1 Gas Turbine
2 Steam Turbine
3 Single Pressure HRSG
4 Supplementary Firing
5 Desalination Unit
6 Feedwater Tank
7 Pressure Reducing Stations

Figure 5-11 Flow Diagram of a Cycle Coupled with a Seawater Desalination Plant

If a high electrical efficiency is required, a combined-cycle is a good solution because electrical efficiencies of up to 50% can be attained with such desalination applications. A conventional steam power plant would usually not reach this level in pure electricity generation mode at the ambient conditions that usually apply. A combined-cycle plant is less suitable if the ratio between fresh water produced and electrical power must be high, i.e., a low power coefficient. In that case, either the additional steam demanded for the desalination unit must be supplied from an auxiliary boiler or a different type of power plant must be chosen.

Repowering

Converting existing steam power plants into combined-cycle plants is known as repowering. It is ideal for plants in which the steam turbines, after many years of operation, still have considerable service-life expectancy, but the boilers are ready for replacement. The boilers are normally replaced or supplemented with gas turbines and HRSGs. Steam turbine units in older power stations generally have relatively low live-steam data and can easily be adapted for use in a combined-cycle. Repowering to a combined-cycle can improve the efficiency of an existing plant to a level close to that of new combined-cycle plants. Some plants are repowered purely in order to benefit from this efficiency increase even though they are far from the end of their design life.

Conversion of conventional steam power plants

Figure 5-12 is a simplified flow diagram for a typical conventional steam turbine plant before repowering. The conventional boiler (1) generates steam at one pressure level for the single-pressure steam turbine (2). There are four extractions from the steam turbine for feed-water preheating–three to LP preheaters (4) and one to the feedwater tank (3). Cooling in this case is done using a water-cooled condenser (5). HP preheaters and reheat are not shown in this example but their presence would not affect the principle of the repowering concept.

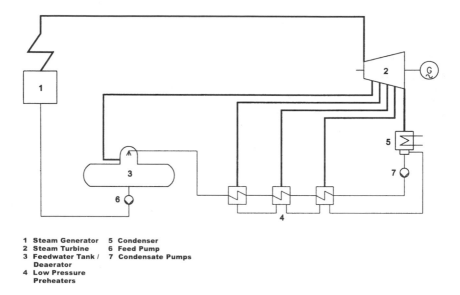

1 Steam Generator   5 Condenser
2 Steam Turbine   6 Feed Pump
3 Feedwater Tank / 7 Condensate Pumps
  Deaerator
4 Low Pressure
  Preheaters

Figure 5-12 Flow Diagram of a Conventional Non-reheat Steam Power Plant

Figure 5-13 shows the same cycle after repowering with an oil-fired gas turbine and single-pressure HRSG. A preheating loop is installed in the HRSG to supply steam for deaeration (6). This means that the existing heaters are not needed and the steam turbine extractions must be blocked off. An additional feedwater pump (15) must be installed for the preheating loop.

For natural gas fuel with a low sulfur content, a dual pressure cycle would have been chosen. This would utilize the exhaust gas even further by adding dual-pressure economizers at the cold end of the HRSG. If the existing feedwater tank cannot be used under vacuum then a condensate-preheating loop would be installed in the HRSG to utilize the exhaust gas energy as much as possible.

If a concept using several pressure levels, or even reheat, is used, the question arises where to feed the steam produced into the existing steam turbine. LP steam can often be admitted into the crossover pipe between the HP and LP parts of the turbine. If the steam turbine is too small, it may not have an internal crossover pipe. In such a case, one of the existing feed heating extractions could be adapted for use as an admission point.

1 Compressor
2 Gas Turbine
3 HP Superheater
4 HP Evaporator
5 HP Economizer
6 LP Evaporator
7 HP Drum
8 LP Drum
9 Steam Turbine
10 HP Steam Bypass
11 LP Steam Bypass
12 Condenser
13 Condensate Pump
14 Feedwater Tank /
   Deaerator
15 LP Feedwater Pump
16 HP Feedwater Pump
17 Pegging Steam Line

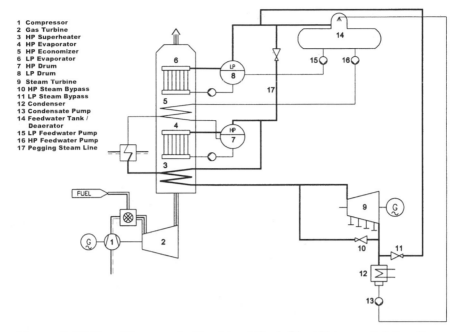

Figure 5-13 Flow Diagram of a Combined-Cycle Plant Using an Existing Steam Turbine

When repowering, a decision must be made about equipment to retain from the existing steam plant. This will vary from case to case and depends on technical and economic criteria. Typically the following will be reused:

- building and foundations
- steam turbine and generator
- condenser
- main cooling system
- main transformer for the steam turbine
- the high-voltage equipment

These larger items would be expensive to replace. On the other hand, it also means that the repowering is designed within any technical limits they might impose. It may however be preferable to replace some of the smaller components of the steam cycle as this can be done at relatively low cost. Sometimes, too, retaining them creates unforeseeable extra costs and may have a negative effect on the availability of the repowered installation. These items include:

- condensate pumps
- feedwater pumps
- control equipment
- piping and fittings
- valves

In order to achieve efficiencies close to those of green field combined-cycles it is important to have a good fit between the size of the gas turbine(s) and the steam turbine. If the steam turbine is too large and the gas turbine is too small, the HRSG will not be able to produce enough steam to build up sufficient steam pressure in the steam turbine to achieve good thermodynamic cycle data. The size relationship between the steam turbine and gas turbines is therefore a main efficiency driver in a repowering application.

Gas Turbine with a Conventional Boiler

Modern steam power plants with reheat steam turbines can be repowered using a concept called the "hot wind box" to improve the efficiency.

Conventional boilers use fresh-air fans to supply the air for combustion. Combustion air can also be provided by the exhaust gas of a gas turbine, if it is installed near the existing steam generator and adapted to accommodate this new operating mode. Because gas turbine exhaust-gas temperature is much higher than fresh air after an air preheater–typically up to 550 to 650°C, (1,022 to 1,202°F) versus 300 to 350°C (572 to 662°F)–modifications are required to the burners, fresh air ducts, and heat transfer sections of the main boiler. Such repowering is highly sophisticated and questionable considering today's low installation cost for new equipment.

Figure 5-14 is the flow diagram for such a cycle. The gas turbine exhaust gas flows first through the conventional boiler (4) and then through a waste-heat recovery system (5 and 8) used for most of the feedwater preheating. The rest of the preheating is done using the

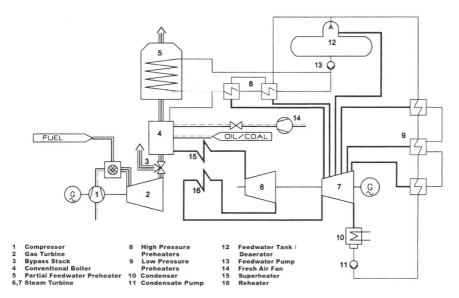

Figure 5-14 Flow Diagram of a Gas Turbine Combined with a Conventional Cycle (Hot Wind Box)

1	Compressor	8	High Pressure	12	Feedwater Tank /
2	Gas Turbine		Preheaters		Deaerator
3	Bypass Stack	9	Low Pressure	13	Feedwater Pump
4	Conventional Boiler		Preheaters	14	Fresh Air Fan
5	Partial Feedwater Preheater	10	Condenser	15	Superheater
6,7	Steam Turbine	11	Condensate Pump	16	Reheater

existing steam turbine extractions and preheaters (8 and 9). The fresh air fan (14) is retained for use in case the gas turbine is out of operation. The bypass stack in the gas turbine exhaust (3) provides extra operating flexibility and allows the gas turbine to operate in single-cycle mode if the water/steam cycle or main boiler is out of operation.

Apart from the complications that may arise in adapting the boiler and the overall operating concept for the new mode of operation, a major factor to consider for this type of repowering is that space must be made available on-site for the gas turbine and the waste-heat recovery system, relatively near to the conventional boiler.

For steam power plants that burn gas or oil, however, the efficiency can be raised by more than 10% and power output by 20 to 30% (relative). With coal-burning units, there is less potential gain because the conversion itself is more complex and there is less improvement in efficiency.

High-Efficiency Coal and Gas Cycle (HEC&G)

Figure 5-15 shows the HEC&G cycle–a repowered cycle designed to have a very high efficiency and operating flexibility.

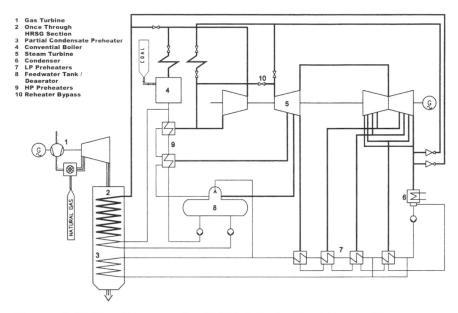

1 Gas Turbine
2 Once Through
   HRSG Section
3 Partial Condensate Preheater
4 Convential Boiler
5 Steam Turbine
6 Condenser
7 LP Preheaters
8 Feedwater Tank /
   Deaerator
9 HP Preheaters
10 Reheater Bypass

Figure 5-15 Flow Diagram of an HEC&G Cycle, Hybrid Power Plant

A special aspect of this concept is that it is primarily intended for conventional plants where the conventional boiler is to remain in operation. The gas turbine (1) and HRSG (2 and 3) are installed in parallel to the conventional boiler to provide a second source of HP live steam for the steam turbine. Again, the exhaust heat at the cold end of the HRSG is used for feedwater preheating. The HP section of the HRSG is designed to generate HP steam in a once-through section that is appropriate for high pressures. The existing steam extractions and feed heaters are reused but only partially fed with water, allowing steam turbine extraction flows to be reduced and increasing steam turbine output.

In Figure 5-15, total live-steam flow from the HRSG and conventional boiler expands in the steam turbine (5) and the total amount of cold reheat steam is fed to the reheat section of the conventional boiler where it is reheated before expanding through the rest of the steam turbine. To maintain maximum efficiency the reheat temperature should be kept at the original level. If the additional flow from the HRSG is relatively high compared to that of the conventional plant, measures must be taken to maintain the reheat temperature (e.g., exhaust gas recirculation, tilting burners, etc.).

There are three main operating modes for this concept:

- original mode without the gas turbine and HRSG in operation
- hybrid mode, where the conventional cycle, gas turbine, and HRSG are in operation at the same time
- combined-cycle mode, where the gas turbine, HRSG, and steam turbine are in operation without the conventional boiler. If the original cycle has reheat, a reheat bypass is necessary for this mode of operation

Table 5.1 shows the some performance data for a converted 500 MW conventional steam turbine power plant with the 178 MW gas turbine.

Table 5-1 Conversion of a 500-MW Steam Turbine Power Plant into an HEC&G: Performance data for different operating modes.

		Original mode	Hybrid mode	Combined-cycle mode
Fuel Input (LHV)	MW	1,312.9	1,496.6	473.0
Steam Turbine Output	MW	502.3	495.4	87.0
Gas Turbine Output	MW	0	178	178
Total Gross Output	MW	502.3	673.4	265.0
Gross Efficiency (LHV)	%	38.26	45.0	56.0
Auxiliary Consumption	MW	32.7	30.5	4.9
Total Net Output	MW	469.6	642.9	260.1
Net Efficiency (LHV)	%	35.77	42.96	55.00
Net Heat Rate (LHV)	kJ/kWh	10,065	8,380	6,545
Net Heat Rate (LHV)	Btu/kWh	9,540	7,943	6,204
Net marginal Efficiency of gas (LHV)	%	-	58.90	55.00

Highest overall efficiencies are obtained with pure combined-cycle mode and the highest output with hybrid mode. However, in hybrid mode the net marginal gas efficiency (the contribution of the gas-fired in the gas turbine to the total efficiency) is higher than in combined-cycle mode. This means that a hybrid plant is a very efficient way of burning natural gas, making this an interesting alternative to separate plants. Combinations of the above modes are also possible resulting in extremely high operating flexibility and a wide operating regime.

The choice of operating mode at any moment will depend on the economics of the plant. As the relative prices of gas and coal and the demand for electrical output vary, the operator can decide which mode can be run most economically at any one time.

Special Applications

There are several variations in the way a gas cycle can be combined with a steam cycle which largely deviate from the type of combined-cycles we have discussed so far. Some of these cycles have proven to be of great interest and others are still being researched and developed. As yet, few of these have achieved any commercial status because demand is limited and the processes and equipment involved can be very complex. Some of the most important of these applications will be discussed here.

Applications with alternative fuels

Where no natural gas is available, there is increasing interest in looking for ways to use available alternative fuels such as coal, heavy oil and crude oils. Often these fuels have a high sulfur content or are unsuitable for firing in gas turbines, so they must be treated before they can be used. Sometimes steam or heating is needed in the treatment process, which can be provided by the combined-cycle plant and results in some very interesting but complex integrated processes. Two of these processes have gained commercial acceptance and are described below.

Integrated gas combined-cycle (IGCC). This consists of a combined-cycle plant integrated with a gasification process. The raw fuels are usually the by-products of an oil refinery, such as heavy oil with high sulfur content, or coal. These fuels are gasified and cleaned to produce a synthetic gas known as syngas that can be fired in a combined-cycle without exceeding permissible emissions levels. Steam is often exported from the combined-cycle to the gasification process and sometimes also to the refinery from which the raw fuel is supplied. There may be a direct supply of auxiliary power to the gasification process as well. Typical efficiencies are only slightly greater than conventional steam turbine power plants and the process is very complex and expensive, but it may be practical where natural gas is unavailable or if natural gas is expensive compared to coal.

Pressurized fluidized-bed combustion (PFBC). These processes are of great interest because they enable coal to be burned cleanly. They are integrated combined-cycles as shown in Figure 5-16. The gas turbine has an inter-cooler (4) between a low-pressure (1) and a high-pressure (2) compressor. At the outlet of the compressor air is fed to the combustion vessel (5) with a pressure of 12 to 16 bar (160 to 218

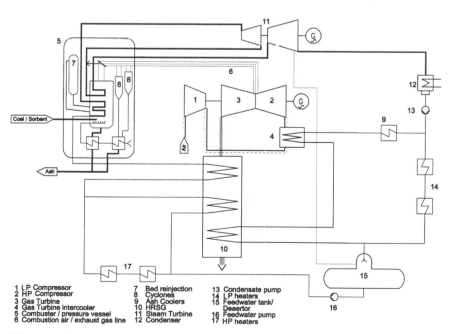

1 LP Compressor	7 Bed reinjection	13 Condensate pump
2 HP Compressor	8 Cyclones	14 LP heaters
3 Gas Turbine	9 Ash Coolers	15 Feedwater tank/
4 Gas Turbine intercooler	10 HRSG	Deaertor
5 Combuster / pressure vessel	11 Steam Turbine	16 Feedwater pump
6 Combustion air / exhaust gas line	12 Condenser	17 HP heaters

Figure 5-16 Flow Diagram of a Simplified PFBC Process

psig) and forced through the fluid bed feeding the combustion process. Coal is mixed with sorbent before entering the combustion process. To obtain a high degree of desulfurization, the bed is maintained at a temperature of 850°C (1,560°F). Before leaving the combustion vessel the exhaust gas is led through cyclones (8) to clean it before it is expanded in the gas turbine (3). The exhaust heat of the gas turbine is recovered in an economizer (10), which is used for the final stages of feed water preheating (17) for the combustor. The steam turbine is a single-pressure turbine with reheat (11).

Net efficiencies are clearly above those of conventional power plants burning coal, although well below the level of pure combined-cycle plants because of the relatively low combustion temperature (i.e., 850°C (1,562°F)).

Another advantage of these cycles is that combustors are very compact, taking up much less space than a conventional boiler, though this is partly counteracted by the large economizer and gas turbine. Also, the small storage capacity of the steam generator means that the steam process reacts very quickly to changes in demand from the control system.

However, again, the design is complex and requires a special type of steam generator/combustor. This means that the investment cost is high compared to the simple combined-cycle plant. Also, no separate operation of the gas turbine and steam turbine is possible.

Steam Injection into the Gas Turbine (STIG Cycles)

We have already seen how injection of steam into the gas turbine can be used to reduce NOx emissions. This decreases the efficiency of the combined-cycle but increases the power output significantly. Cycles in which generated steam is injected into the gas turbine are known as STIG (steam-injected gas-turbine) cycles.

Figure 5-17 shows a STIG cycle in which all the generated steam is directed into the gas turbine (4), apart from a small amount used for feed water preheating and deaeration. Since steam is only required in the gas turbine at one pressure level, a single-pressure HRSG is used (2). The quantity of make-up water to the cycle is very high because all of the generated steam is lost to the atmosphere in the

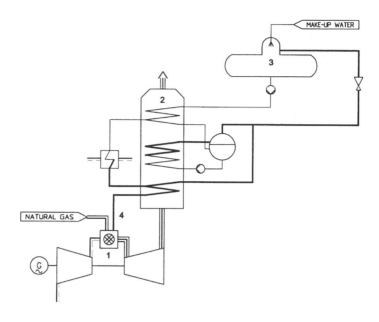

1 Gas Turbine
2 HRSG
3 Feedwater Tank / Deaerator
4 Steam Injection into Combuster

Figure 5-17 Flow Diagram of a STIG Cycle

process. This type of cycle is suitable for use as a peaking unit in countries where water is plentiful. It is simple, attaining a high specific output and a higher efficiency than the gas turbine alone, though not as high as that of a combined-cycle. However, if the injected steam flow is equal to more than approximately 2% to 4% of the air-mass flow, major modifications must be made to the gas turbine. This limits the commercial viability of the concept because of the fact that gas turbines are standardized and are only modified for special applications unless these applications are widely used.

Another, more efficient variation on this type of cycle which consumes less water is the Turbo-STIG cycle, shown in Figure 5-18. This is suitable for smaller cogeneration plants with aero-derivative gas turbines. A steam turbine (4) is installed on the same shaft as the gas turbine (1) and shares the gas turbine generator. The single-pressure HRSG (2) generates steam for the steam turbine, expanding it to a level suitable for reheating in the HRSG and injecting (6) into the gas turbine. Clearly, the steam turbine adds to the power output and effi-

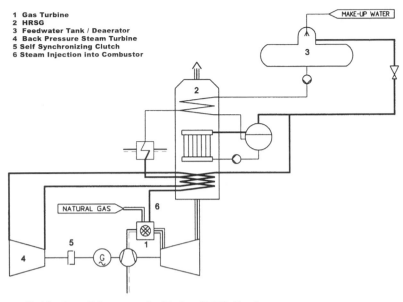

1 Gas Turbine
2 HRSG
3 Feedwater Tank / Deaerator
4 Back Pressure Steam Turbine
5 Self Synchronizing Clutch
6 Steam Injection into Combustor

MAKE-UP WATER

NATURAL GAS

Figure 5-18 Flow Diagram of a Turbo-STIG Cycle

ciency but still does not reach the level of a normal combined-cycle. This, and the fact that the water consumption is so high, is part of the reason for the limited commercial acceptance of the STIG cycle.

Humid Air Turbine (HAT Cycle)

The principle here is to increase gas turbine output by using humid air as the working fluid. A flow diagram is shown in Figure 5-19. Heat from compressor cooling (3 and 4) and the gas turbine exhaust is used to generate steam sufficient to result in an air flow with up to 25% steam content. This air is then heated, again using gas turbine waste heat, before entering the turbine.

Efficiencies reached with this cycle are not as high as those of normal combined-cycles but the HAT cycle can reach higher specific output levels. The cycle is also suited for use with coal gasification plants where available low-grade energy could be used to generate the required steam.

Difficulties in the HAT cycle lie in the operational complexity and the high water consumption that would cause problems where water is scarce. Also, new gas turbine development is needed due to the changed flow pattern in both the compressor and turbine of the gas turbine. This will probably be the main obstacle in commercializing this cycle.

Cycles with Alternative Working Media

A less-proven way of improving the performance of a combined-cycle plant is by using fluids other than pure water/steam in the bottoming cycle. The idea is that by using a mixture of fluids, such as water mixed with ammonia, the evaporation in the HRSG will no longer take place at only one temperature but over a range of temperatures. This serves to reduce the exergy loss between exhaust gas and working fluid thereby increasing the efficiency.

The cycles operate at only one pressure and are more suitable for gas turbines with relatively low exhaust temperatures such as aero-

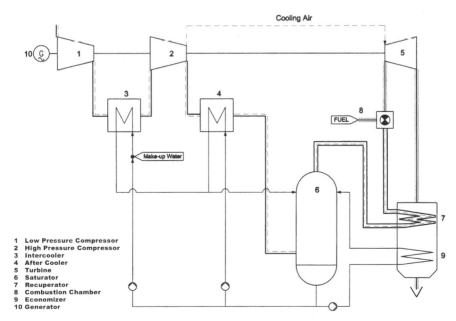

Figure 5-19 Flow Diagram of a HAT Cycle

derivative machines. Condensation takes place at more than one temperature, which complicates the condensing system. There is a further environmental disadvantage, in that leakages could be very dangerous, because the fluids in question are aggressive.

As gas turbines develop towards higher inlet and exhaust gas temperatures the advantages of these alternative working media will vanish.

COMPONENTS

CHAPTER 6

6

Components

Gas Turbine

The gas turbine is the key component of the combined-cycle plant generating approximately two-thirds of the total output.

The gas turbine process is simple: ambient air is filtered, compressed to a pressure of 14 to 30 bar (190 to 420 psig), and used to burn the fuel producing a hot gas with a temperature generally higher than 1,000°C (1,832°F). This expands in a turbine driving the compressor and generator. The expanded hot gas leaves the turbine at ambient pressure and at a temperature between 450 to 650°C (842 to 1,202°F) depending on the gas turbine efficiency, pressure ratio, and turbine inlet temperature.

The combined-cycle plant has become a competitive thermal power plant only as a result of the rapid development in gas turbine technology, which is still ongoing. Gas turbine development tends towards increasing gas turbine inlet temperatures (e.g., by improving cooling technologies) and increasing compressor air flows.

Increasing gas turbine inlet temperatures produces a higher useful enthalpy drop and therefore increases the efficiency and output of the gas turbine and combined-cycle plant. This can usually be done using the same compressor and even though further investment in materials may be necessary the specific cost of the plant is reduced. Since fuel costs and capital costs are the main drivers for the cost of electricity the gas turbine inlet temperature should be increased to the potential of the material to improve the competitiveness of the product.

Parallel to the development of the turbine there has also been some development in the compressor. Today's compressors can handle much larger air mass flows and higher pressure ratios, resulting in higher power outputs, reduced specific costs, and improved efficiency.

Additionally, competitiveness is increased with new gas turbine concepts, such as sequential combustion that appeared on the market in the 1940s and was re-introduced in the 1990s. The gas turbine has two stages of combustion, with an intermediate turbine section, and a higher exhaust-gas temperature. This means higher combined-cycle efficiencies can be achieved without raising the firing temperatures.

Figure 6-1 shows the historical development of maximum air flows and gas turbine inlet temperatures.

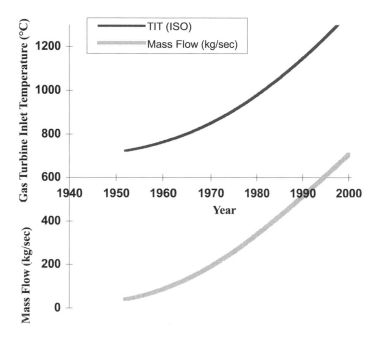

TIT: Turbine Inlet Temperature

Figure 6-1 Historical Trends in Gas Turbine Inlet Temperatures and Compressor Air Flows

Two Categories of Turbines

Power generating gas turbines can be classified into two categories:

- the aeroderivative gas turbine, consisting of a jet turbine modified for industrial duty and frequently incorporating a separate power turbine
- heavy duty industrial gas turbines (originally derived from steam turbine or jet technology)

The *aeroderivative gas turbine* is normally a two- or three-shaft turbine with a variable-speed compressor and driving turbine. This is an advantage for part-load efficiency since airflow is reduced with the lower speed.

Inlet temperatures in jet engines are higher than in industrial gas turbines. In a jet turbine, too, weight plays the dominant role and so product and maintenance costs are less important than with industrial gas turbines, where long intervals between inspections are demanded. For that reason, and because of smaller dimensions in the hot gas path, the trend toward higher inlet temperatures and greater power densities has progressed more rapidly in jet turbines than in stationary machines.

Aeroderivatives generally offer higher efficiencies than their frame counterparts as a result of aerotechnology. Furthermore, they are smaller and lighter for a given power output and can be started rapidly because of their inherent low inertia. Because these turbines are derived from jet engines they retain many of the features designed to allow rapid "on the wing" maintenance of aero engines. The small physical size of the aeroderivative enables it to be removed from its installation and replaced within a day.

A disadvantage, when the aeroderivative is operating with a generator, is that there is no compressor braking the power turbine during load shedding. Two shaft turbines are usually used for compressor or pump drives, where the operating speed of the power turbine is also variable. Due to the size of aeroplanes, aero derivatives are limited to approximately 50 MW electrical output.

Heavy duty industrial gas turbines are practically always built as single-shaft machines when used to drive a generator with an output greater than approximately 30 MW. Due to the progress with turbine inlet temperatures and compressor air-mass flows, gas turbines for power generation can today achieve up to 300 MW electrical output.

Figures 6-2 and 6-3 show typical modern gas turbines. Figure 6-2 shows an aeroderivative unit and Figure 6-3 a gas turbine designed for 160 MW output originally derived from steam turbine technology.

As gas turbines have become standardized, they have begun to be manufactured on the basis of sales forecasts rather than order intake. This results in a series of frame sizes for each gas turbine manufacturer and hence shorter installation times, lower costs, and lower

New
compressor
replaces fan

LP bleed added
for low speed
operation

Dry lowemissions
combustor replaces
aero annular can

Last two stages
of LPT and exhaust
redesigned

Rear drive
added

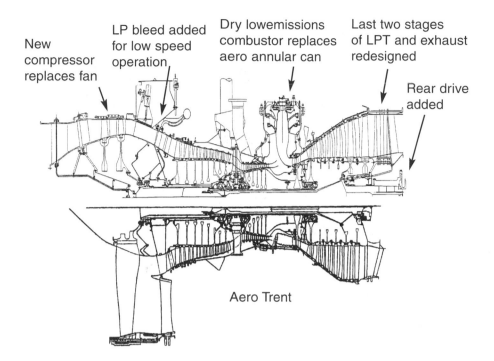

Aero Trent

Figure 6-2 Industrial Trent Derived from the Aero Trent 800. Source: Rolls-Royce plc. Registered trademark Rolls-Royce plc.

prices. Major developments in heavy duty industrial gas turbines have taken place in the last decade. A major impact has been made by the introduction to the market of gas turbines with sequential combustion.

Gas Turbines with Sequential Combustion

In a gas turbine with sequential combustion, air enters the first combustion chamber after the compressor. Here, fuel is burned, raising the gas temperature to the inlet temperature for the first turbine. The hot gas expands through this turbine stage, generating power before entering the second combustion chamber, where additional fuel is burned to reach the inlet temperature for the second part of the turbine, where the hot gas is expanded to atmospheric pressure.

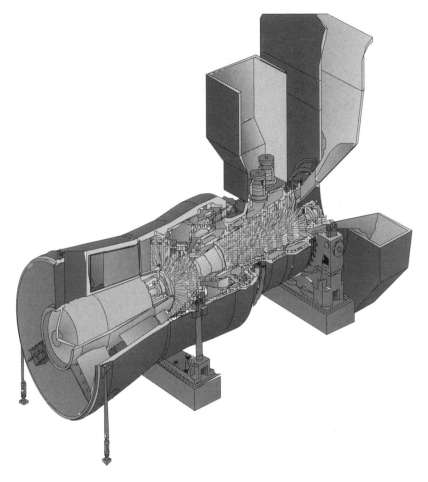

Figure 6-3 Heavy Duty Industrial Gas Turbine

With the same gas turbine inlet temperatures as in a gas turbine with a single combustion a higher efficiency can be achieved, with the same emission levels.

Figure 6-4 shows a gas turbine with sequential combustion. The compressor is designed for a pressure ratio of 30, followed by the high-pressure combustion chamber, the high-pressure turbine, low-pressure combustion chamber, and low-pressure turbine. A compact arrangement fits all this equipment into one casing on a single shaft.

Table 6-1 indicates the main characteristic data of modern gas turbines designed for power generation.

Table 6-1 Main Characteristic Data of Modern Gas Turbines for Power Generation

Power output (ISO condition)	Up to 300 MW
Efficiency (ISO condition)	34-40%
Gas turbine inlet temperature (ISO 2314)	1,100-1,350°C (2,012-2,462°F)
Exhaust gas temperature	450-650°C (842-1,202°F)
Exhaust gas flow	50-550 kg/s (397,000-4,370,000lb/hr)

In the past, gas turbines were mainly developed for use in simple-cycle operation and the majority of gas turbine installations were of this type. These gas turbines were often "peaking" and stand-by machines. The first combined-cycle power plants were based on the same gas turbines but operated in intermediate and baseload applications.

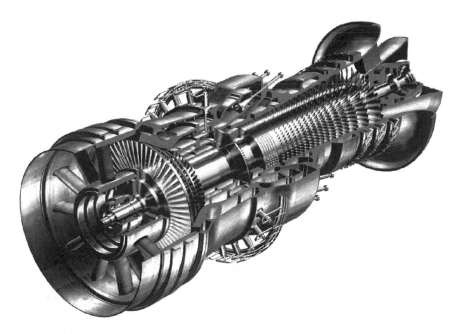

Figure 6-4 Gas Turbine with Sequential Combustion

Increasingly, the main application for the gas turbine is in combined-cycle, which has a corresponding impact on gas turbine development and equipment. The main market breakthrough of the gas turbine in combined-cycle operation took place in the mid-1980's and it is now a mature, widely applied technology. Improving gas turbine performance lowers the cost of electricity, increases the competitiveness of the product, and so motivates manufacturers to invest in new types and upgrades.

Often, because new models of given gas turbines are sold before the first unit running has accumulated many operating hours, attention must be paid to the corresponding risk mitigation.

Inspections

Gas turbines achieve a very high level of reliability through regular and properly administered inspection and maintenance activities. In a combined-cycle plant, the inspections for the other equipment can mostly be done in the window given by the gas turbine inspection schedule.

Degradation

The output of a gas turbine is subject to some degradation for two main reasons:

- *fouling*, which is recoverable. During operation, attention must be paid to compressor and turbine fouling
- *aging*, which is non-recoverable unless parts are replaced

Compressor fouling occurs because the gas turbine operates with an open-air cycle. The compressor is fed with air that cannot be cleaned completely. Compressor fouling is reduced by an air filtration system that is suited to the environment at the plant site. The filters most frequently used are two-stage filters or self-cleaning pulse filters. The second of these is only suitable for use in dry climates.

It is, however, impossible to keep the compressor completely clean. The fouling which results causes losses in output and efficiency that are greater in simple-cycle gas turbines than in combined-cycle plants. This is due to the fact that in combined-cycle plants, some of the losses can be recovered in the steam cycle.

Two types of compressor cleaning can be used to help recover these losses:

- on-line washing
- off-line washing

On-line washing is straight forward but because the temperature increases through the compressor, the solution evaporates and cleaning is limited to the first compressor rows.

Off-line washing is better suited to large, modern gas turbines because it is more effective. However, it requires shutting down and cooling the engine. Washing at low speed is preferred (e.g., ignition speed with the machine cold). The machine is therefore out of operation for approximately 24 hours, mainly for cooling off and for drying the engine after washing. This type of washing is best done before or after an inspection. The amount of cleaning solution needed is in the range of 40-200 liters per washing cycle.

After compressor cleaning of a gas fired machine, any remaining degradation is due to aging. The symptoms of aging after 8,000 hours in operation on a clean fuel are:

- reduction in output from the combined-cycle plant of about 0.8 to 1.5%
- reduction in efficiency of the combined-cycle plant of about 0.5 to 0.8%

The rate of aging decreases with time between major overhauls.

Turbine fouling is caused principally by the ash contained in heavy fuels and in the additives used to inhibit high-temperature corrosion. It is unavoidable with heavy fuels, but can be limited by selecting the proper type of additives. Heavy and crude oil fired gas turbines can also be equipped with a turbine washing system. Turbine fouling is less of a problem in peaking or intermediate-load operation because of the self-cleaning effect produced by start-up and shut down.

Typical degradation (after compressor washing) in combined-cycle plants after 8,000 hours of operation on heavy or crude oil:

- reduction in output of the combined-cycle plant of 4 to 5.5%
- reduction in efficiency of the combined-cycle plant of 1.5 to 1.9%

The rate of degradation also decreases with time.

In the past, corrosion problems were one of the major causes of gas turbine failure. Because of the use of better blading materials and coatings, problems of this nature have practically been solved. Whenever heavy fuels are used, particularly those containing vanadium or sodium, it is necessary to use additives or treat the fuel to prevent high-temperature corrosion. The additives commonly used are based on magnesium, chromium, or silicon.

Heat Recovery Steam Generator

The heat recovery steam generator (HRSG) is the link between the gas turbine and the steam turbine process. There are three main categories:

- HRSG without supplementary firing
- HRSG with supplementary firing
- steam generators with maximum supplementary firing

As can be seen from chapter 4, HRSGs without supplementary firing are most common in combined cycle plants.

The function of the HRSG is to convert the exhaust energy of the gas turbine into steam. After heating in the economizer, water enters the drum, slightly subcooled. From the drum, it is circulated to the evaporator and returns as a water/steam mixture to the drum where water and steam are separated. The saturated steam leaves the drum for the superheater where it reaches the maximum heat exchange temperature with the hottest exhaust-gas leaving the gas turbine. The heat exchange in an HRSG can take place on up to three pressure levels depending on the desired amount of energy and exergy to be recovered.

Today, two or three pressure levels of steam generation are most commonly used.

HRSG without supplementary firing

Construction. An HRSG without supplementary firing is essentially an entirely convective heat exchanger. The requirements imposed by the operation of the combined-cycle power plant are often underestimated. In particular, provision must be made to accommodate the short start-up time of the gas turbine and the requirements for quick load changes. The HRSG must be designed for high reliability and availability. HRSGs can be built in two basic constructions, based on the direction of gas turbine exhaust flow through the boiler.

Vertical HRSG. In the past, vertical HRSGs were most often known as forced- circulation HRSGs because of the use of circulating pumps to provide positive circulation of boiler water through the evaporator sections. In this type of boiler, the heat transfer tubes are horizontal, suspended from uncooled tube supports located in the gas path. Vertical HRSGs can also be designed with evaporators that function without the use of circulating pumps.

Figure 6-5 shows a forced circulation HRSG. The exhaust-gas flow is vertical, with horizontal tube bundles suspended in the steel structure. Circulating pumps assure constant circulation within the evaporator. The structural steel frame of the steam generator supports the drums.

Horizontal HRSG. The horizontal type of HRSG has typically been known as the natural-circulation HRSG because circulation through the evaporator takes place entirely by gravity, based on the density difference of water and boiling water mixtures. In this type of boiler, the heat transfer tubes are vertical, and essentially self-supporting.

Figure 6-6 shows a natural circulation HRSG. The exhaust-gas flow is horizontal. The steel structure is more compact than on a unit with vertical gas flow.

Design Comparison. Either type–vertical or horizontal–can be used in a combined-cycle plant. In the past, vertical HRSGs had sev-

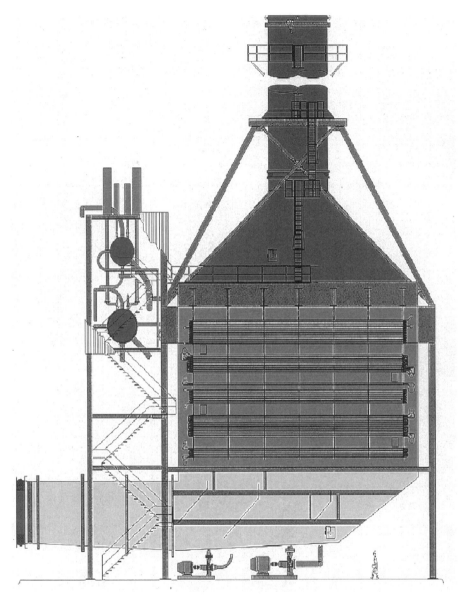

Figure 6-5 Forced Circulation Heat Recovery Steam Generator

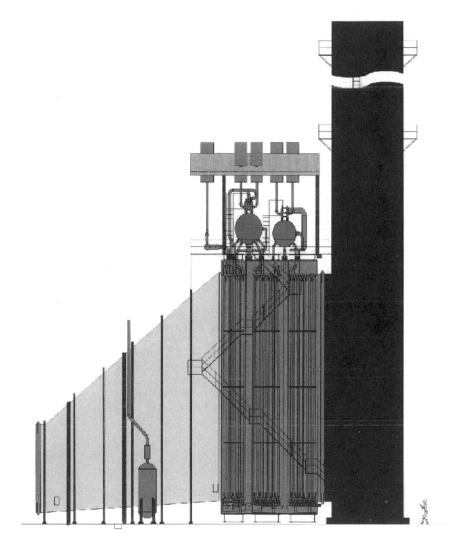

Figure 6-6 Natural Circulation Heat Recovery Steam Generator

eral advantages that made them especially suited for combined-cycle applications:

- minimum space requirements arising from the vertical design
- smaller boiler volumes because of the use of smaller diameter tubes
- less sensitivity to steam blockage in economizers during startup

The main advantage of the horizontal HRSG is that no circulation pumps are needed–an important point for applications with design pressures above 100 bar (1,430 psig), where pumps must be designed and operated with special care. Additionally, there is an advantage with vertical tubes in the evaporator since the tubes with the highest heat absorption in the evaporator have the most vigorous circulation, and tube dry-out can not occur in vertical tubes.

Current design for natural-circulation boilers has overcome the disadvantages relative to vertical boilers. Space requirements and start-up times are identical, water volumes in evaporators have been reduced with use of smaller diameter tubes, and steam blockage is better handled in a modern, natural-circulation boiler. The same pinch-points can be achieved in the high- and intermediate-pressure evaporator. Only in large HRSGs with a tight low pressure pinch-point do differences occur in steam performances.

The presence of both technologies on the market indicates that both meet customer expectations and the preference for one or the other is more on a historical or on a regional preference (e.g., in North America, the overwhelming majority of HRSGs are of natural circulation design).

Once-through HRSG. The HRSGs described above use a steam drum for water/steam separation and water retention. Combined-cycle plants are often operated in cycling duties with frequent load changes and start-stop cycles. HRSGs employing a drum with design pressures of 100 bar (1,430 psig) and beyond impose restriction on this operation.

In a once-through steam generator, the economizer, evaporator, and superheater are basically one tube–water enters at one end and steam leaves at the other end, eliminating the drum and the circulation pumps. This design has advantages at higher steam pressures since the drum does not limit start-ups and load changes. Both horizontal and vertical HRSGs can be built with the once-through circulation principle.

Figure 6-7 shows a once-through HRSG with horizontal gas flow. The high-pressure portion is of the once- through type, while the low-pressure portion is of conventional, natural-circulation design with an LP drum on top of the boiler.

Figure 6-7 Once-Through HRSG with a Drum Type LP Section

Casing of the HRSG. The casing for the HRSG can be categorized in two main designs:

- a cold-casing design with inner insulation
- a hot-casing design with external insulation

The cold-casing design is widely used with natural circulation steam generators and has advantages at high exhaust-gas temperatures. The casing withstands the exhaust-gas pressure forces and the inner insulation keeps it at low temperature. The casing construction also imposes no limits on start-up times because of thermal expansion of the casing.

The hot casing design is often used with the vertical HRSGs and has advantages at lower exhaust-gas temperatures. When fuels with a high sulfur content are fired, a hot casing at the cold end of the HRSG can limit corrosion.

With modern gas turbines running with high exhaust temperatures, a hot-casing design requires high-alloy materials in the hot end of the boiler. Today, some vertical HRSGs also use a cold-casing design.

Finned Tubing. Heat transfer in the HRSG is mainly by convection. Heat transfer on the water side is much better than on the exhaust-gas side, and so finned tubes are employed on the exhaust-gas side to increase the heat-transfer surface.

Normal fin density for an HRSG behind a gas turbine (firing natural gas or No. 2 oil) is 5 to 7 fins per inch (200 to 280 fins per meter). If the gas turbine uses heavier fuels the fin density is reduced to 3 to 4 fins per inch (120 to 160 fins per meter) to better handle deposits.

The optimum HRSG must fulfill the following, sometimes contradictory conditions:

- the rate of heat recovery must be high (high-efficiency)
- pressure losses on the exhaust-gas side of the gas turbine must be low in order to limit losses in power output and efficiency of the gas turbine
- the permissible pressure gradient during start-up must be large

- low-temperature corrosion must be prevented

It is particularly difficult to meet the first two conditions at the same time. Because of the relatively low temperature, the heat transfer takes place mainly by means of convection. The differences in temperature between the exhaust-gas and the water (or steam) must be small in order to obtain a good rate of heat recovery, which requires large surfaces. This means large pressure losses in the exhaust-gas unless the velocity of the gas is kept low, which would result in a lower heat transfer coefficient and therefore further increase surfaces. Using small-diameter finned tubes helps to solve this problem.

Another benefit of small-tube diameters is the small amount of water in the evaporator, resulting in a smaller thermal constant, which favors quick load changes.

HRSGs being built today have low pinch-points and small pressure drops on the exhaust-gas side. Pinch-point values of 8-15 K (14-27°R) at pressure losses of 25 to 30 mbar (10 to 12" w.c.) are attainable even with a triple-pressure reheat boiler.

Low-temperature corrosion. When designing HRSGs, care must be taken to prevent or restrict low-temperature corrosion. To accomplish this, all surfaces that come into contact with the exhaust-gas must be at a temperature above or slightly below the sulfuric acid dewpoint. When burning a sulfur-free fuel in the gas turbine the limit is determined by the water dewpoint.

Suitable precautions enable operation of the heat exchangers at temperatures below the acid or water dewpoints. This can be done by selecting appropriate materials or by adding corrosion allowances to the affected tubes. Since there are only a few tubes in this temperature range it can be beneficial despite the low exergy of the heat gained.

Optimum design of an HRSG. Designing an HRSG involves optimizing between cost and gain. The main cost driver is the heat exchanger surface installed. The indicator generally used is the pinch-point in the evaporator. The surface of the evaporator increases exponentially as the pinch point decreases while the increase in steam generation is only linear. For that reason the pinch-point selected is a critical factor in determining the heat transfer surface. In installations where efficiency is highly valued, the pinch-point is 8 to 15 K (14 to

27°R); where efficiency is of lower value, it can be higher, 15 to 25 K (27 to 45°R).

Operating experience. A challenge affecting the design of the HRSG is the quick start-up of the gas turbine, especially for cold and warm plant conditions. The rapid thermal expansions that occur during a start-up can be accommodated through suitable design measures such as suspension of the tube bundles, drum design, tube to header connections, etc.

The main constraint for the loading rate often arises from the drum. To make as quick a start as possible, the walls of the drum should be as thin as possible, which can be done provided the design live-steam pressure is sufficiently low. Modern gas turbines provide higher exhaust-gas temperatures than older models, therefore higher live-steam pressures, especially with reheat steam cycles are now more attractive resulting in longer start-up times.

Once-through HRSGs eliminate the thick high-pressure drum and therefore provide the desired high thermal flexibility.

Another point of concern is the volumetric change within the evaporator during start-up. The large differences in specific volume between steam and water at low and intermediate pressures cause large amounts of water to be expelled from the evaporator at the start of the evaporation process. If the drum cannot accommodate most of this water, a great amount of water would be lost through the emergency drain of the drum during every start-up, or an undesired emergency trip of the unit would be required to avoid water carryover into the steam system.

To improve part-load efficiency and behavior of the combined-cycle plant, the boiler is operated in the sliding-pressure mode. That means the system is generally operated at a lower pressure when the steam turbine is not at full load. This can be accomplished by keeping the steam turbine control valves fully open. For example, in a system with two gas turbines and two HRSGs feeding a common steam turbine, half-load of the whole power station can be accomplished with only one of the gas turbines running at full load. In sliding pressure operation the live-steam pressure is at 50% of the pressure at full load. The steam volumes in the evaporator, superheater and live-steam lines of the HRSG in operation are doubled.

During off-design conditions, economizers can start to generate steam which can block tubes and reduce the performance of the HRSG. In order to keep this within limits, the economizer is dimensioned so that the feedwater at the outlet is slightly subcooled at full load. This difference, between the saturation temperature and the water temperature at the economizer outlet, is known as the approach temperature. Because it causes a reduction in the amount of steam generated, it should be kept as small as possible (typically 5 to 12 K (9 to 22°R)). Proper routing of the economizer outlet tubes to the drum also prevents blockage if there is steaming in the economizer.

Another way of preventing steaming in the economizer is to install the feedwater control valve downstream of the economizer, the economizer is kept at a higher pressure and steaming is prevented.

HRSG with limited supplementary firing

Despite the fact that the majority of the HRSGs are unfired, there are occasional applications where a limited amount of supplementary firing is needed.

The operating principles of an HRSG with limited supplementary firing are the same as those for the unfired boiler. There are various designs available for the firing itself. Units that do not exceed a gas temperature of approximately 780°C (1,436°F) after the supplementary firing can be built with simple duct burners. This limit can be extended but requires modifications to the design of the HRSG.

Figure 6-8 shows an HRSG with supplementary firing. This system is particularly well-suited to burning natural gas, which attains uniform temperature distribution after the burners. Radiation to the walls of the combustion chamber is relatively low. For that reason, the majority of the HRSGs of this type burn natural gas. There are systems available for burning oil but the burner equipment is more complex and costly to install, operate and maintain.

The importance of supplementary firing in HRSGs for power generation alone is diminishing. This is mainly caused by two facts:

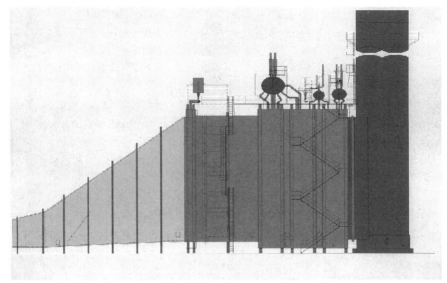

Figure 6-8 Supplementary Fired Heat Recovery Steam Generator

- modern gas turbines have exhaust-gas temperatures closer to the maximum allowable HRSG inlet gas temperature reducing the effect of any supplementary firing
- supplementary firing behind a modern gas turbine results in an efficiency decrease for the combined-cycle. With older gas turbine models the efficiency was more or less constant

Supplementary firing is most often applied in combined-cycle cogeneration plants where the amounts of process steam must be varied independently of the electric power generated. In this case, supplementary firing controls the amount of process steam generated. Additional applications include burning gases which are not suitable for firing in the gas turbine (e.g., pressure is too low or heat content unsuitable) and reaching a higher power output, but with the expensive penalty of a lower efficiency of power generation.

A larger amount of fuel could be burned if the combustion chamber of the duct burner had cooled walls. Circulating water from the evaporator provides cooling. Despite the higher operating flexibility, the additional cost and complexity very seldom justify this solution.

Arranging supplementary firing among different heat exchanger bundles can also increase HRSG output. In this application, the exhaust-gas is cooled after the gas turbine by a heat exchanger such as an evaporator or superheater. The exhaust-gases can then be reheated in a duct burner before passing through the remaining portion of the HRSG. This solution is rare and reserved for niche applications that have difficulty competing with an unfired HRSG or other solutions.

Steam Generator with Maximum Supplementary Firing

The maximum firing rate is set by the oxygen content of the gas turbine exhaust-gases. With this type of steam generator, the exhaust-gases of the gas turbine are used primarily as oxygen carriers. The heat content of the exhaust-gas of the gas turbine is small compared with the heat input of the firing in the boiler. It is therefore no longer correct to speak of an HRSG.

The design of a steam generator of this type is practically identical to that of a conventional boiler with a furnace, except that there is no regenerative air preheater. The gas turbine exhausts at a temperature of 450 to 650°C (842 to 1,202°F), rendering a regenerative heater unnecessary. In order to make it possible to cool the exhaust-gas to a low temperature after the steam generator, an additional economizer is provided which takes over a portion of the feedwater preheating from the regenerative preheating of the steam turbine. The best arrangement divides the feedwater between the economizer and the high-pressure feedheaters. When the fuel is gas, an additional low-pressure partflow economizer improves efficiency. The fuel burned in the boiler may be oil, gas, or pulverized coal.

This application can be used to increase the output from an existing conventional steam turbine plant using a gas turbine and its exhaust energy. Due to the high integration of both cycles (e.g., the burners of the existing boiler must be retrofitted to make them suitable for the exhaust-gas application), the complexity and the number of interfaces are considerable.

Steam Turbine Technology

The most important requirements for a modern combined-cycle steam turbine are:

- high efficiency
- short start-up times
- short installation times
- floor-mounted installation

In the past, combined-cycle steam turbines were applications of industrial steam turbines or derivatives from conventional steam turbine plants. The main differences between combined-cycle steam turbines and industrial steam turbines are:

- larger power output
- higher live-steam temperatures and pressures
- fewer applications of extractions
- new installations frequently have reheat steam turbines

The main differences between combined-cycle steam turbines and conventional steam turbines are:

- fewer or even no bleed-points as opposed to 6 to 8 for feed-heating
- floor-mounted arrangement
- shorter start-up times
- smaller power output
- lower live-steam pressures 100 to 160 bar (1,430-2,310 psig) as opposed to 160 to 300 bar (2,310-4,340 psig)

With the different requirement profile and the big volume of the combined-cycle plant orders, an optimized design for this application is justified.

Characteristics of Combined-Cycle Steam Turbines

Steam turbines used for combined-cycle installations are simple machines. In the past they used relatively low live-steam data; with the raising of the gas turbine exhaust temperatures, optimal steam pressures increased. Live-steam temperatures have now reached those of the conventional steam power plants as well.

Combined-cycle plants frequently generate steam at more than one pressure level. Due to multiple inlets, the steam-mass flow in the steam turbine increases from the inlet towards the exhaust (the bleed for the partial feedwater preheating gives only a small reduction). In a conventional steam turbine, the steam-mass flow is reduced to roughly 60% of the inlet flow at the exhaust.

Short start-up times are of particular importance because combined-cycle plants are often used as medium-load units with daily or weekly start-ups and shut-downs.

Table 6-2 illustrates how in recent years, the combined-cycle plant has changed gas turbine technology, producing a big impact on steam turbine requirements:

Table 6-2 Change of Boundary Conditions for Steam Turbines in Combined-Cycle Plants

	Traditional	**Modern**
Gas turbine exhaust temperature	450-550°C (842-1,022°F)	550-650°C (1,022-1,202°F)
Live steam temperatures	420-520°C (788-968°F)	520-565°C (968-1,049°F)
Number of steam pressure levels	1 or 2	2 or 3
Live steam pressure	30-100 bar (420-1,430 psig)	100-160 bar (1,430-2,310 psig)
Reheat steam cycle	no	Yes
Number of gas turbines per steam turbine	1 to 5	1 or 2

Multi- and Single-Shaft Plants

In a multi-shaft combined-cycle plant, there are generally several gas turbines with HRSGs generating steam for a single steam turbine. The steam and gas turbines use different shafts, generators, step-up transformers, etc. By combining the steam production of all the HRSGs, a larger steam volume enters the steam turbine, which generally raises the steam turbine efficiency.

Modern gas turbines are larger in output with high exhaust temperatures. With the largest gas turbines on the market, one steam turbine per gas turbine or one steam turbine for two gas turbines is common.

If one steam turbine per gas turbine is installed, the single-shaft application is the most common solution–gas turbine and steam turbine driving the same generator.

A plant with two gas turbines can be built either in a two-gas-turbine-on-one-steam-turbine configuration (multi-shaft) or as a plant with two gas turbines in a single-shaft application. The following differences result:

- The plant with the single-shaft configuration employs the larger generators and has fewer interfaces to the grid (two generators vs. three), resulting in a simpler high-voltage system
- The water/steam cycles in a single-shaft plant are completely separated, which simplifies operation especially for reheat plants
- Gas turbines in the same plant are installed in a staggered sequence for a single-shaft plant. This can be used to achieve earlier commercial operation with the first single-shaft unit while the multi-shaft plant needs both gas turbines to commission the steam turbine. Single-shaft plants enable the owner to extend the plant in smaller units
- Single-shaft power plants have better part-load efficiency because at 50% load only one unit will be in operation with full-load efficiency, whereas the multi-shaft efficiency is lower as the plant is in part load

There are two concepts for single-shaft plants. First, the generator is between the gas turbine and the steam turbine, each turbine driving one end of the generator, the steam turbine engages and disengages with a self-shifting and synchronizing clutch. Second, the generator is at one end, driven by both turbines from the same side. The steam turbine therefore is rigidly coupled to the gas turbine on one side and the generator on the other.

The advantages of having the generator between the gas and steam turbine are:

- the gas turbine can be run up and loaded independently of the steam turbine
- the plant can continue to run for a number of steam turbine failures; the steam produced is dumped into the condenser
- shorter start-up times of the plant or lower power consumption at start-up
- the steam turbine is at standstill during start-up of the gas turbine and does not need cooling steam for start-up
- the flexibility is increased with a self-synchronizing clutch with very high reliability

For single-shaft applications the generator between the gas and steam turbine is designed to be slid out sideways, thus allowing removal of the rotor. The advantages of having the generator at the end is that the cost of the clutch can be saved.

Figure 6-9 shows a steam turbine for a single-shaft unit in a triple-pressure reheat plant. The steam turbine output is 140 MW with live-steam conditions at 110 bar (1,580 psig) and 562°C (1,044°F).

Live-Steam Pressure

Early gas turbines were smaller in power output and with their lower exhaust temperatures, optimal pressure for the steam cycle was low. This was beneficial for the steam turbine for two reasons: Low pressures give larger steam volumes that can be more efficiently expanded in the steam turbine, and with a non-reheat cycle, a low live-

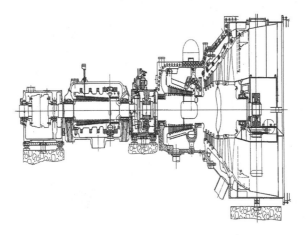

Figure 6-9 Cross Section of a 140 MW Reheat Steam Turbine with a Separate HP Turbine and a Combined IP/LP Turbine with Axial Exhaust

steam pressure limits the moisture content in the steam turbine exhaust and therefore, erosion.

Modern gas turbines with their high exhaust temperatures result in combined-cycles with high live-steam pressures in the steam cycle.

A geared high-pressure steam turbine enables high-efficiency expansion with small live-steam volumes by using a gear to increase the speed of the high-pressure turbine, increasing blade length, reducing secondary blade losses and resulting in an improved overall efficiency.

Figure 6-10 shows a steam turbine with a geared high-pressure turbine that takes advantage of high-steam pressure/small steam volume in a combined-cycle plant. It also shows the small dimensions of the geared high-pressure turbine. The additional investment for the gear is also justified by the increased thermal flexibility of the high-pressure turbine, which is important for the cycling duties of the combined-cycle plant.

Generators

The majority of gas and steam turbines are directly coupled to two-pole generators. For units with ratings below 40 MW, four-pole generators that run at half speed are more economical. It is advanta-

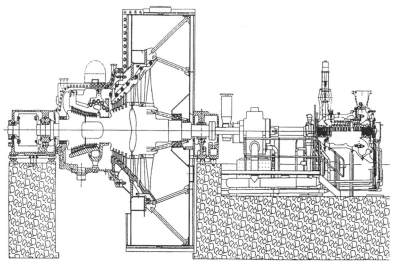

Figure 6-10 Cross Section of a Two Casing Steam Turbine with Geared HP Turbine

geous that turbines of this output usually already have a gear so to apply a four-pole generator, only the reduction ratio of the gear must be adapted.

Three types of generators are used in combined-cycle plants:

- air-cooled generators with a closed-air circuit (totally enclosed water to air-cooled, TEWAC)
- hydrogen-cooled generators
- air-cooled generators with open-air cooling

Generators with *open-air circuit cooling* are low-cost and have no need for additional cooling, but problems with fouling, corrosive atmosphere and noise can arise. Generators with *closed-air circuit cooling* are built for capacities up to 480 MVA. These machines are reasonable in cost and provide excellent reliability. The full-load efficiency of modern air cooled generators is beyond 98%. *Hydrogen-cooled generators* attain an efficiency of approximately 99% at full load, making their efficiency superior, particularly at part-load, to air-cooled machines. However, they require additional auxiliaries and monitoring equipment, are more complex in design, and, as a result, more expensive. In general, they do not reach the high reliability of the closed-air circuit cooled generators.

Figure 6-11 shows a generator with closed-air cooling. Water is used to cool the air. These machines are well suited for use with single shaft combined-cycle plants where the gas turbine and the steam turbine drive the same generator. In plant condition this generator can transmit 320 MVA. Nowadays air-cooled generators of up to 480 MVA have now been built and tested due to the overall benefits of air cooling. For higher outputs hydrogen cooling has to be selected.

Electrical Equipment

The single line diagram for a combined-cycle power plant with a single shaft is shown in Figure 6-12. It is similar to a single line diagram for other types of power stations. Key features in this diagram are:

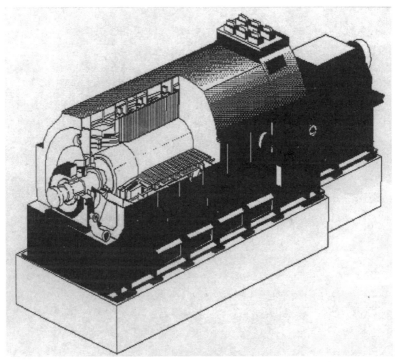

Figure 6-11 Cut-away Drawing of an Air Cooled Generator for use in Combined-Cycle Power Plant

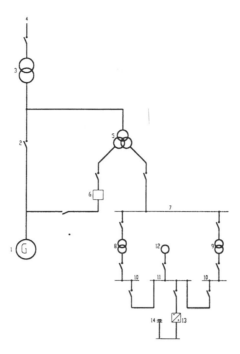

1 Generator
2 Generator breaker
3 Step-up transformer
4 Link to the high voltage grid
5 Station service transformer
6 Static starting device
7 Medium voltage switchgear

8 Auxiliary transformer
9 Auxiliary transformer
10 Low voltage switchgear
11 Emergency switchgear
12 Stand-by diesel set
13 Battery charger
14 Batteries

Figure 6-12 Single Line Diagram

- one generator for the gas and steam turbine (only one outlet to high voltage switchyard needed)
- auxiliaries are fed from the generator output, which saves a separate in-feed from the grid and allows the plant to run at no load during a grid blackout, ready for being reconnected
- three-winding-station service transformer to feed the auxiliaries and to feed the static frequency converter to start the gas turbine using the generator as a motor
- medium voltage bus for large motors
- uninterruptible power supply for important consumers

Control System

The control system is the brain of the power plant. It supervises, controls, and protects the plant, enabling safe and reliable operation.

The gas turbine is supplied with a standardized control system that provides fully automated machine operation. The water/steam process is correspondingly automated to achieve a degree of uniform operation of the plant as a whole, thereby reducing the risk of human error.

For this reason, the control and automation systems of a combined-cycle plant form a relatively complex system even though the thermal process is fairly simple. Fully electronic control systems are applied in modern combined-cycle plants. The main features of the control system for combined-cycle power plants are:

- truly distributed architecture
- complete range of functions for process control
- communication capability due to several bus levels
- compliance with standard communication protocols
- openness for third-party applications
- on-line programmability with easy creation/editing of the programs

A hierarchic and decentralized structure for the open- and closed-loop control systems is best adapted to the logic of the whole process. It simplifies planning and raises the availability of the plant. A highly automated control system encompasses the following hierarchic levels:

At the *drive level*, all individual drives are monitored and controlled. Drive protection and interlockings are provided by the control system. Signal exchange with higher hierarchical levels is also provided.

At the *functional group level*, individual drives for one complete portion of the process are assembled into functional groups. The logical control circuit on this level encompasses interlocks, automatic switching, and preselection of drives. Examples of typical function groups are feedwater pumps, cooling water pumps, and lube-oil systems.

The *machine level* includes the logical control circuits that link the function groups to each other. These include, for example, the fully automatic starting equipment for the gas turbine or the steam turbine.

The *unit level* includes the controls that coordinate the operation of the whole combined-cycle power plant. Figure 6-13 shows these levels in the automation hierarchy in a combined-cycle plant.

For baseload power plants a lower degree of automation can be selected. The frequent starting and stopping of an intermediate- or peak-load-duty plant requires a higher degree of automation.

Process computers provide sequencing events, keep long term records and statistics, provide management information regarding

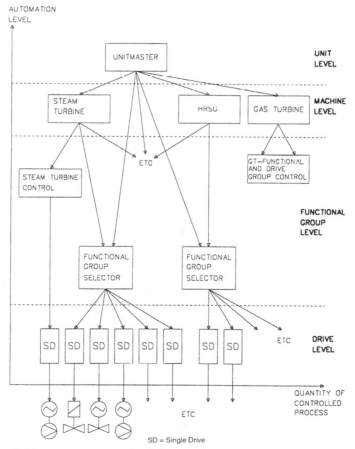

Figure 6-13 Hierarchic Levels of Automation

economy of the plant, optimize the heat rate and operation of the plant, and advise on the intervals between cleanings, inspections and other maintenance work.

Figure 6-14 shows the control room of a combined-cycle plant.

Cooling System

Three different cooling systems can be applied to combined-cycle plants:

- direct air-cooling in an air cooled condenser
- indirect air-cooling with a wet or hybrid cooling tower
- direct water-cooling using river or sea water

For *direct air-cooling*, no cooling water is required, but the output and efficiency of the plant are reduced due to higher vacuum levels. This is used in regions where no water is available or as a way to minimise the impact on the infrastructure and environment thereby facilitating permitting.

Figure 6-14 Standard Layout for a Modern Combined-Cycle Power Plant Control Room

Indirect air-cooling requires water to replace evaporation and blow-down losses. This amount of water depends on the exhaust steam flow of the steam turbine and amounts to approximately 0.3kg/s (2,400 lb/h) per MW installed plant capacity for a combined-cycle plant without supplementary firing.

The *direct water-cooling* variant requires water in an amount 40-60 times larger than for indirect cooling. After serving as heat sink, this water is returned to the water source (e.g., river, sea, or cooling pond) from which it came.

A combined-cycle plant of a size equal to a conventional steam power plant has a steam turbine only one-third of the size and requires only half of the cooling (see chapter 9). This gain in power output and efficiency with a direct-cooling variant over a cooling tower are therefore often not justified due to the high additional cost of civil structures needed for cooling water inlet and outlet.

Bypass Stack

Sometimes combined-cycle plants are equipped with exhaust-gas bypass dampers and bypass stacks for simple-cycle operation of the gas turbines. With modern gas turbines this feature loses importance. The reasons are:

- the design is expensive with gas turbine exhaust temperatures of 600°C and beyond
- if the damper fails, the whole plant is out of service
- an exhaust-gas damper has leakage losses with resultant losses in output and efficiency
- the costs incurred are significant

An exhaust-gas bypass stack (with or without damper) is a prerequisite for phased installation of a combined-cycle plant–one in which the gas turbine runs in simple-cycle operation before the steam cycle is connected. This allows the plant operator to get two-thirds of the final plant capacity on line sooner and earn early income. This can be economical for plants with large capacity payments. However, to connect the steam cycle normal gas turbine operation must be interrupted because the gas turbines are needed for the commissioning of the steam cycle.

In deregulated markets, none of this is justified–the investment for the bypass stack, the expensive power generation during gas turbine simple-cycle operation, and the business interruption to connect the steam cycle–especially since a steam cycle can be installed in less than two years.

Other Components

In addition to the major equipment described above, the combined-cycle plant includes other equipment and systems. For example:

- fuel supply systems including a gas compressor if required
- steam turbine condenser
- feedwater tank/deaerator
- feedwater pumps
- condensate pumps
- piping and fittings
- evacuation system
- water treatment plant
- compressed air supply
- steam turbine bypass
- civil works

These are similar whether used in combined-cycle plants or in other types of power plants and therefore are not described here in detail.

CONTROL AND AUTOMATION

CHAPTER 7

7

Control and Automation

Power plants are normally operated to meet a demand dictated by the electrical grid, and where there is one, the process energy system. The control systems discussed below are used to fulfill these requirements, to ensure correct control in transient modes such as start-up and shut-down, and to assure the safety of the plant under all operating conditions.

One type of control in a combined cycle plant is closed loop control. This is when a controller receives actual measured data as an input, and uses it to correct a signal to a control device, with the aim of reaching the set point of a given parameter. The important closed control loops for a combined cycle fall into two main groups,

- the main plant load control loop
- the secondary control loops which maintain the important process parameters, such as levels, temperatures, and pressures within permissible limits

A third group of closed control loops are component specific, such as minimum flow control for a pump or lube oil pressure control, and are intended primarily for the safe working of individual items of equipment. Since they do not directly affect the control of the cycle as a whole they are not discussed here.

Applied logically, this leads to the hierarchical structure of the entire control system already mentioned in chapter 6.

Load Control and Frequency Response

The electrical output of a combined cycle plant without supplementary firing is controlled by means of the gas turbine only. The steam turbine will always follow the gas turbine by generating power with whatever steam is available from the HRSG.

The gas turbine output is controlled by a combination of variable inlet guide vane (VIGV) control, and gas turbine inlet temperature (TIT) control. The TIT is controlled by a combination of the fuel flow admitted to the combustor and the VIGV setting. Modern gas turbines are equipped with up to three rows of VIGVs allowing a high gas turbine exhaust gas temperature down to approximately 40% load. Below that level, the turbine inlet temperature is further reduced as the airflow cannot be further reduced.

After a gas turbine load change the steam turbine load will adjust automatically with a few minutes delay dependent on the response time of the HRSG. It is, however, sometimes suggested that independent load /frequency control of the steam turbine should be provided for sudden increases or decreases in load. Such a system would require the steam turbine to be operated with continuous throttle control, resulting in much poorer efficiencies at full and part loads and additional complications. Since the gas turbine generates approximately two-thirds of the total power output, a solution without control for the steam turbine power output is generally preferred. This is also supported by the fact, that modern gas turbines react extremely quickly to frequency variations and they can usually compensate for the delay in the steam turbine response with falling frequencies.

Figure 7-1 illustrates the concept of closed loop load/frequency control without separate steam turbine control for a plant with two gas turbines and one steam turbine. An overall plant set point, Ps, is given to the overall plant load control system, KAR, which determines how the load should be distributed between the gas turbines. It receives a load / frequency signal LFC, from each of the three generators which it uses to determine whether it is necessary to make any correction to the load. The load of the individual gas turbines is controlled by setting the position of the VIGVs using inlet guide vane control VIGV, and the gas turbine inlet temperature TIT control which varies the air and fuel flow to the gas turbine. Since the TIT cannot be directly measured, readings are taken of the turbine pressure ratio and exhaust gas temperature TAT, from which the TIT is calculated.

The entire steam cycle is operated in sliding pressure mode with fully open steam turbine valves down to approximately 50% live steam pressure. This is the mode of operation best suited for high part load efficiencies. More sophisticated control is not absolutely necessary be-

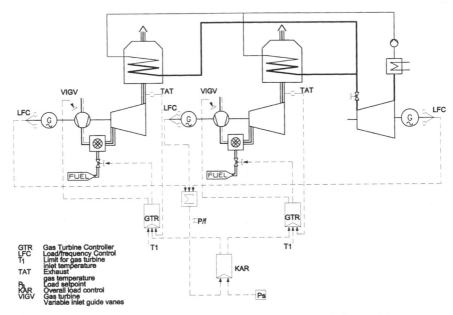

GTR Gas Turbine Controller
LFC Load/frequency Control
T₁   Limit for gas turbine
     inlet temperature
TAT Exhaust
     gas temperature
Pₛ   Load setpoint
KAR Overall load control
VIGV Gas turbine
     Variable inlet guide vanes

Figure 7-1 Principle Diagram for a Combined-Cycle Load Control System

cause the power output of the plant can be adjusted by changing the set points of the individual gas turbine controls.

If supplementary firing is provided, it may be beneficial to provide independent load control for the steam turbine. The steam process then operates in a manner similar to that of a conventional steam plant, where the amount of steam generated is varied to fit demand by adjusting the supplementary firing fuel input.

Up to this point, no distinction has been made between load and frequency control. In principle, the remarks remain valid for both. However, another very important aspect of the load/frequency control is the ability of a plant to react to rapid fluctuations in frequency that may occur in the electrical grid. This is known as frequency response, and must normally be done in a matter of seconds, whereas loading takes place over several minutes.

In order to sustain stable operation of a plant, a grid frequency dead band of typically +/- 0.1 Hz is introduced within which the plant will not respond. Outside this dead band a droop setting is followed. The standard gas turbine droop setting is 5%, which means that a grid frequency drop of 5% would cause a 100% load increase. The droop characteristic setting is defined during the planning phase, and is typically in the range of 3 to 8%.

For plant configurations without steam turbine load control the steam turbine will not be able to support falling frequencies, within the 10 to 15 seconds normally required by grid codes, so the total response will have to come from the gas turbine. However, for increasing frequencies both the gas turbine and steam turbine will be able to support the grid. In such a mode the steam turbine valves will just close allowing less steam to expand through the steam turbine.

Figure 7-2 shows a typical droop characteristic for a combined cycle power plant with a 5% droop setting. If the frequency drops by 1%, the gas turbine load will initially jump 20% and the steam turbine load 0%. This result corresponds to a combined cycle response of approximately 13%. On the other hand if the frequency increases by 1% the combined cycle response of 20% is achieved by load changes of 20% in both the gas turbine and steam turbine, illustrated in the diagram by equivalent gas turbine load change of 27%.

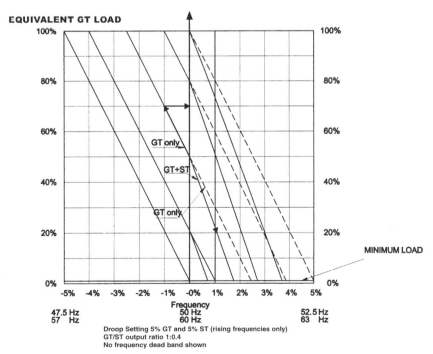

Figure 7-2 Typical Combined Cycle Droop Characteristic

In order to perform a plant load jump while the frequency is falling, it is essential that the gas turbine is operating below the maximum output level. Any operation, even in frequency support mode, above this level maximum is not possible. For frequency support gas turbines are typically operated between 40 and 95% load.

In conclusion, it can be said that combined cycle plants are very well suited to rapid load changes. Gas turbines react extremely quickly because their time constant is very low. As soon as the fuel valve opens, more added power becomes available on the shaft. Gas turbine load jumps of up to 35% are possible, but they are not recommended because they are very detrimental to the life expectancy of the turbine blading.

Secondary Closed Control Loops

Figure 7-3 shows the essential closed control loops that are required to maintain safe operating conditions in a dual pressure combined-cycle. The diagram shows the input signals into each controller and the command signals to the control valves. These control loops, which are typical for all combined-cycle concepts are described below :

Drum level control, BLC-HP and BLC-LP

This is normally a three-element control system, as shown, which forms one signal from the feed water and live steam flows, and the level within the drum. This signal is then used to position the feed water control valve.

Live steam temperature control

The live steam temperature control loop in drum type HRSGs actually limits rather than controls the live steam temperature. Its purpose is to reduce the temperature peaks during off-design operating conditions such as hot ambient temperatures, part load and peak load. For that reason, attemperation often takes place after the superheater and not between two portions of the superheater as in a conventional

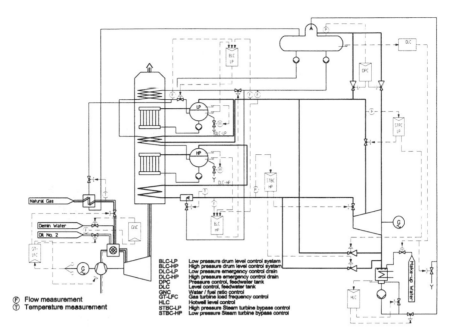

Figure 7-3 Closed Control Loops in a Combined-Cycle Plant

steam generator. Normally, high pressure feed water is injected into the live steam to cool it down to the required temperature.

No real temperature control extending over a broad load range is possible in purely HRSG operation because the turbine exhaust gas temperatures drop off with cold ambient temperatures and extreme part loads of the gas turbine. In plants with supplementary firing, the relationships are more like those of a conventional steam generator. Because elevated gas temperatures are possible in these cases, it is important that the temperatures of the steam and the superheater tubes be maintained within safe limits. To do this, the superheater could be divided into two sections with the attemporator installed in between them.

If the HRSG is of the once-through type, the live steam temperature is generally controlled by the feed water flow to the HRSG and the gas turbine exhaust gas temperature. With such an arrangement, however, attemperation is still required at extreme part load conditions and during start-up.

Feed water temperature, DPC

In order to avoid low temperature corrosion, the feed water temperature must not, even in the lower load range, drop significantly below the acid or water dew point. On the other hand, this temperature must be as low as possible in order to assure good utilization of the heat available, as discussed in chapter 4. It is therefore recommended that the feed water temperature be held at a constant level corresponding approximately to the acid dew point.

For the control loop shown on the diagram deaerator heating is done primarily with a steam turbine extraction, but if the pressure in this extraction is not sufficient at any load point then pegging steam from a live steam source (LP in figure 7-3) is used. In plants with a low pressure preheater loop in the HRSG, the opposite problem could occur during part load operation of the gas turbines, where more low pressure steam could be generated than required. This excess energy must be dissipated either by increasing the pressure in the feedwater tank/deaerator, which reduces the amount of steam generated, or by directing the excess low pressure steam to the condenser, which again slightly increases the vacuum and therefore marginally lowers the steam turbine output.

For plants without a feedwater tank the minimum feed water temperature is normally controlled by hot water recirculation from the low pressure economizer outlet into the HRSG feed water inlet, see also Figure 4-45. To compensate for the pressure losses in the economizer a small booster pump is foreseen. The amount of water is determined by the desired feed water temperature. For high sulphur fuels, normally back-up fuels, the economizer is by-passed and the water enters directly into the low pressure drum.

Live steam pressure, STBC-HP and STBC-LP

Usually the steam turbine is in sliding pressure operation down to approximately 50%, so continuous control for the live steam pressure is superfluous in this range. Below this the pressure is kept constant by closing the steam turbine valves. Control is, however necessary for non-steady state conditions such as start-up, shut-down, or malfunction.

How this control is accomplished depends on the plant equipment. The live steam pressure is controlled by the steam turbine control valves and/or steam turbine bypass control valves.

A steam bypass provides the following advantages:
- flexible operation during start-up, shut-down, turbine trip, or quick changes in load
- shorter start-up times
- environmental acceptability (since no steam is vented to the atmosphere)

Table 7-1 shows when main steam turbine valves and steam bypass valves, are used for various modes of operation of a combined cycle plant with an unfired HRSG.

Table 7-1 Operation of Steam Turbine Control Loops in a Combined-Cycle Plant

Type of operation	Steam by-pass	ST control valves
Start-up	+	+
Shut-down	+	+
Normal sliding pressure operation	-	-
Fixed pressure operation below 50%	-	+
Steam turbine switch off or trip	+	-
HRSG switch-off	+	-
Gas turbine trip	+	-

+ in operation, - not in operation

Level in feedwater tank and hotwell, DLC and HLC

These levels ensure that there is sufficient head for the feedwater and condensate pumps and the necessary water buffer in the cycle. Hotwell level is controlled by adjusting the main condensate flow control valve, which will open for increasing and close for decreasing hotwell levels. If the level in the feed water tank is too high the drain valve in the main condensate line is opened to prevent the level rising further. If the feedwater tank level is too low make-up water is admitted to the cycle, usually via the condenser, which in turn raises the hotwell level and thereby the feed water tank level.

Supplementary firing

If the plant has supplementary firing, this will usually be controlled independently to meet a required output or steam demand, by regulating the amount of fuel admitted to the burners. This is often used to control variations in process steam flows, which cannot be achieved by the combined cycle alone. If demand falls, the supplementary firing load is normally reduced first before the gas turbine load is decreased.

Process energy

If there is a process extraction it is usually controlled independently of the main process control loops described above. Sometimes the control may even be done externally to the combined cycle, in the plant that is receiving the process. Usually, however, the pressure must be regulated, either internally or externally to the steam turbine, or using a combination of both. Sometimes the temperature must be controlled with attemporation.

Start-Up and Shut-Down of the Combined-Cycle Plant

Combined-cycle power plants are usually operated automatically. It must therefore be possible to activate equipment during start-up and shut-down from the central control room. Whether the commands are to be issued to the individual drives or drive groups by the operating staff or from a higher level automatic starting program must be decided on a case-by-case basis. In base-load installations operating with few starts, full automation of the steam process is not always necessary.

The dynamic behavior of modern combined cycle plants is characterized by the short start-up time and quick load change capability. Above all, the gas turbine can be started and loaded quickly. Because its reaction time is also short, it is capable of following quick changes and surges in load.

Modern combined cycle plants in the 50 to 400 MW range can be started within the following times:

- hot start after 8 hours standstill: 40 to 50 minutes
- warm start after 60 hours standstill: 75 to 110 minutes
- cold start after 120 hours standstill: 75 to 150 minutes

Gas turbine start-up is independent of the standstill time, allowing two-thirds of the power to be available within 30 minutes after activating the start-up sequence. The start up time for the steam turbine depends on the time required to heat parts of the machine, without exceeding thermal stress limits imposed by the material. Therefore the stand still time is very significant, because it determines what temperature these parts are at when the startup sequence is initiated.

The combined cycle start-up procedure is divided into three main phases:

- purging of the HRSG
- gas turbine speed-up, synchronization and loading
- steam turbine speed-up, synchronization and loading

In order to prevent the explosion of any unburned hydrocarbons left in the system from earlier operation, it may be advisable, especially when firing oil, to purge the boiler before igniting the gas turbine. This is done by running the gas turbine at ignition speed (approximately 30% of nominal speed) with the generator as a motor, or with similar starting equipment, in order to blow air through the

HRSG. The purge time depends on the volume behind the gas turbine e.g. by-pass stack or HRSG, which has to be exchanged by up to a factor of five with "clean air" before ignition of the gas turbine can take place.

After purging, the gas turbine is ignited and run up to nominal speed, synchronized and loaded to the desired load.

During gas turbine start-up, depending on the actual start-up condition (hot, warm or cold) steam is generated more or less rapidly in the HRSG. In general, appropriate steam properties for a steam turbine start-up are reached at approximately 50 to 60% gas turbine load. This means 40 to 60% of nominal pressure and a sufficient degree of superheat (i.e., around 50 K (90°R)). Before starting the steam turbine the gland steam system must be in operation and the condenser evacuated. Until the steam turbine takes over the available steam flow, the excess steam flows across the steam turbine bypass. If supplementary firing has been provided, it should not be brought into operation until the gas turbine is at full load, the steam turbine bypasses are closed, and the steam turbine can accommodate the additional steam flow. With the supplementary firing on, the steam process can then be further loaded.

Figures 7-4 to 7-6 show the three different start-up sequences for a 250 MW class combined cycle plant, which can be brought to full load in as little as 43 minutes after 8 hours standstill. Start- up without purging is assumed.

In Figure 7-6, for starting after 120 hours standstill, a hold point of the gas turbine is introduced. This means that the gas turbine load is held constant, in this case at 15%, while the steam turbine is in the early stages of loading. Therefore the majority of the steam produced in the HRSG can be fed to the steam turbine and must not be dumped to the condenser resulting in a more fuel economic start-up. After the hold point the gas turbine and steam turbine are both run up to full load, the gas turbine reaching full load about thirty minutes before the steam turbine.

Figure 7-7 illustrates how the plant is shut down by reducing the gas turbine load. Once the exhaust gas temperature has reached a prescribed minimum level, the steam turbine is shut down. The boiler and the gas turbine are then further unloaded and shut off.

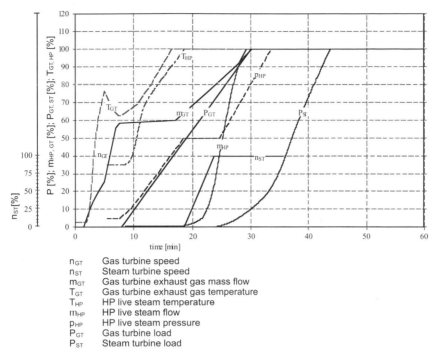

n_{GT}	Gas turbine speed
n_{ST}	Steam turbine speed
m_{GT}	Gas turbine exhaust gas mass flow
T_{GT}	Gas turbine exhaust gas temperature
T_{HP}	HP live steam temperature
m_{HP}	HP live steam flow
p_{HP}	HP live steam pressure
P_{GT}	Gas turbine load
P_{ST}	Steam turbine load

Figure 7-4 Start-up Curve for a 250 MW Class Combined-Cycle after 8 h Stand still

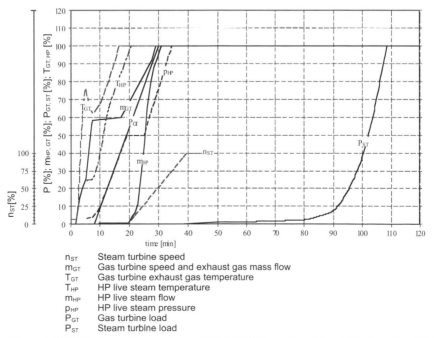

n_{ST}	Steam turbine speed
m_{GT}	Gas turbine speed and exhaust gas mass flow
T_{GT}	Gas turbine exhaust gas temperature
T_{HP}	HP live steam temperature
m_{HP}	HP live steam flow
p_{HP}	HP live steam pressure
P_{GT}	Gas turbine load
P_{ST}	Steam turbine load

Figure 7-5 Start-up Curve for a 250 MW Class Combined-Cycle after 60 h Standstill

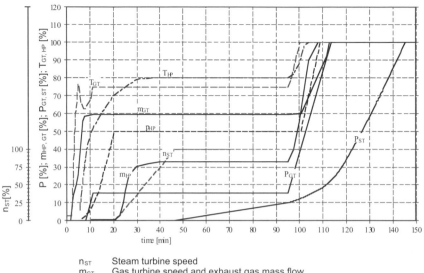

n_{ST}	Steam turbine speed
m_{GT}	Gas turbine speed and exhaust gas mass flow
T_{GT}	Gas turbine exhaust gas temperature
T_{HP}	HP live steam temperature
m_{HP}	HP live steam flow
p_{HP}	HP live steam pressure
P_{GT}	Gas turbine load
P_{ST}	Steam turbine load

Figure 7-6 Start-up Curve for a 250 MW Class Combined-Cycle after 120 h Standstill

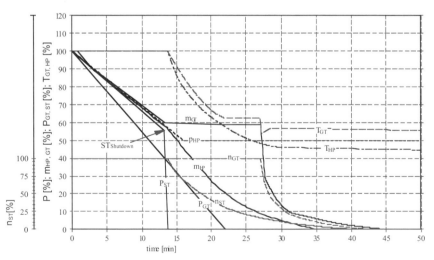

n_{GT}	Gas turbine speed
n_{ST}	Steam turbine speed
m_{GT}	Gas turbine exhaust gas mass flow
T_{GT}	Gas turbine exhaust gas temperature
T_{HP}	HP live steam temperature
m_{HP}	HP live steam flow
p_{HP}	HP live steam pressure
P_{GT}	Gas turbine load
P_{ST}	Steam turbine load

Figure 7-7 Combined-Cycle Shut Down Curve

OPERATING AND
PART LOAD BEHAVIOR

CHAPTER 8

8

Operating and
Part Load Behavior

The way in which a power plant responds to changes in its operating environment is of great importance for the overall plant economy. The changes are normally imposed by different ambient conditions and the power demanded by the grid. In order to maximize the profit for the owner it is necessary to maintain a low cost of electricity throughout the whole operating range.

One component of the cost of electricity is the fuel cost which depends on the plant efficiency. The plant-operating concept will determine at which operating points the plant should have the best efficiencies. Normally baseload efficiency makes a major contribution, and part-load efficiency has to be considered, as it is always important for periods of low demand or for plants continuously operating at part-load to support the grid with spinning reserve.

Combined-Cycle Off-Design Behavior

As opposed to the plant design—in which all components of the water/steam cycle are defined to meet certain design criteria—the performance of a plant under off-design conditions will depend on the behavior of these fixed components with changes in the operating environment.

To find the best possible plant for design and off-design conditions, it is important to know both the steady-state and the dynamic operating behavior of the plant throughout the whole operating range. Theoretical calculations of the dynamic behavior require precise knowledge of all plant components. For that reason, to predict this behavior, operating experience from other similar plants (or sometimes a simulator) is frequently used, combined with steady-state plant calculations. This is an accurate approach, because a dynamic response always leads to a steady-state operating point, if the necessary time factor is considered.

Gas turbine calculations are straightforward since gas turbines are standardized machines and calculation tools and correction curves are available for different ambient conditions and part-load points. The water/steam cycle of the combined-cycle is normally calculated by referring all values to the thermodynamic data at the design point. If that design point is known, general equations such as the Law of Cones and the heat transfer law can be used to reduce the calculation to a reasonable number of equations without the need to consider such things as the dimensions and geometry of the components. A brief study of the details of the method of calculation can be found in the Appendix.

Combined-cycle steady-state behavior differs significantly from that of conventional steam plant. The differences involve mainly the boiler and the operating mode of the plant. In an HRSG, the heat is transferred mainly by means of convection, while in a conventional boiler the main mode of heat transfer is radiation. Additionally, the steam turbine of a combined-cycle plant operates most economically in sliding-pressure mode (i.e., it is run "uncontrolled" and the steam data are determined by a combination of the gas turbine exhaust gas properties and the swallowing capacity of the steam turbine). By contrast, a conventional plant is often operated at a fixed pressure and with a constant live-steam temperature. That simplifies calculations because the steam pressure and temperature are known in advance. The steam turbine and the boiler can therefore be considered independently of one another.

Combined-Cycle Off-Design Corrections

Variations in the behavior of a given combined-cycle due to off-design conditions are very similar to the influences in which the conditions for the combined-cycle power plant design point are varied. This is mainly because gas turbines–which account for two-thirds of the power output at the design point–are standardized machines, and are not redesigned for specific applications. They respond in the same way, therefore, whether the variation in conditions is due to a new design point or to an off-design case for the combined-cycle. The resulting power output and exhaust gas properties to the HRSG will be the same.

The impact on the water/steam cycle is different for off-design and design corrections, however it influences only one-third of the total power produced.

Operating behavior will differ from plant to plant depending on the actual design point. In the following section, some typical variations are shown. These parameters should normally be considered when correcting the combined-cycle performance from one set of operating conditions to another (e.g., for combined-cycle performance testing).

- Plant load
- Ambient air temperature
- Ambient air pressure
- Ambient relative humidity
- Cooling water temperature (not necessary if re-cooled by air)
- Frequency
- Power factor and voltage of the generators
- Process energy
- Fuel type and quality

Effect of Ambient Air Temperature

At the design point, ambient air temperature has a large influence on the power output of the gas turbine and the combined-cycle plant. Behavior of the off-design plant under changes in ambient air temperature is very similar to that for different design points.

In particular, efficiency increases slightly here too if the air temperature increases while the vacuum within the condenser remains constant (refer to Fig. 4-53). However, this effect is marginal and mainly affects the terminal point temperature differences in the HRSG and condenser, giving changes in performance so slight, they are hardly detectable on the curve, compared to the design influence discussed in chapter 4.

Effect of the Ambient Pressure

The main factor influencing ambient pressure is the site elevation. This is purely a design issue, as described in chapter 4. For a given site the power plant may see daily weather variations which are the only cause of change in the ambient pressure. These corrections are basically the same as those described in chapter 4. The correction will only affect plant output and the efficiency will remain constant.

Effect of the Ambient Relative Humidity

The effect of changes in the ambient relative humidity is also similar to that described in chapter 4. An increase in relative humidity increases the enthalpy of the working media of the cycle. It gives more energy to the HRSG than at the design point, causing slightly higher energy transfer through the HRSG sections. This leads to a marginal increase in the terminal point temperature differences of the HRSG in order to transfer this additional energy, giving a slight negative tendency for off design calculations.

Cooling Water Temperature

A change in cooling water temperature affects the volume flow of the steam turbine exhaust steam. The steam turbine exhaust area is selected at the design point. Away from this point the exhaust steam volume flow will be different. This increases the exhaust losses when the cooling water temperature falls so the benefit due to a better condenser vacuum is reduced. If the cooling water temperature is higher, the condenser pressure increases, thereby reducing the steam turbine output. The operating behavior is thus quite different from that which would arise if the size of the turbine were in all cases adapted to the temperature of the cooling water.

Figure 8-1 shows the effect of the steam turbine back-pressure on the relative efficiency of the combined-cycle plant for a typical direct-cooling and a wet-cooling tower application. It is a plant-specific

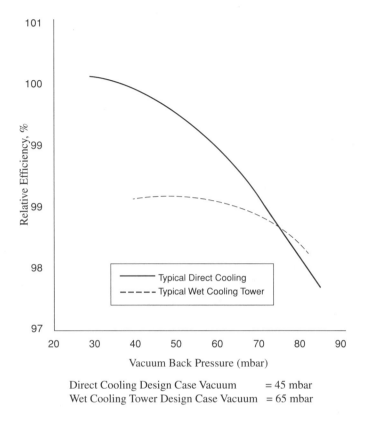

Figure 8-1 Effect of Vacuum on Combined-Cycle Efficiency

curve based on a given steam turbine with a fixed exhaust area that was designed at a certain vacuum. The curve is therefore only generally applicable for plants with the identical design conditions. To determine the condenser back-pressure of different cooling media see Figure 4-58 which is still valid as long as the plant runs at full load.

For part-load conditions, less steam is produced in the HRSG, reducing the amount of steam leaving the steam turbine exhaust to the condenser. If the cooling water flow is maintained for full as well as part loads, the vacuum is reduced even further for part loads, due to the lower heating of the cooling media and the smaller terminal point temperature difference of the condenser.

Electrical Corrections

Frequency: The grid frequency has a major impact on plant behavior, as it determines generator speed and therefore gas turbine speed. The gas turbine compressor's speed defines the airflow entering the gas turbine, which is significant for plant performance. Gas turbines are normally designed to operate at nominal firing temperatures for frequencies from 47.5 to 52.5 Hz for a 50-Hz grid, and 57 to 63 for a 60-Hz grid. The same design criteria are valid for steam turbines.

Figure 8-2 shows a typical variation of combined-cycle output and efficiency for frequency variations. The output decreases for falling frequencies and the efficiency stays within a narrow range of the nominal frequency point.

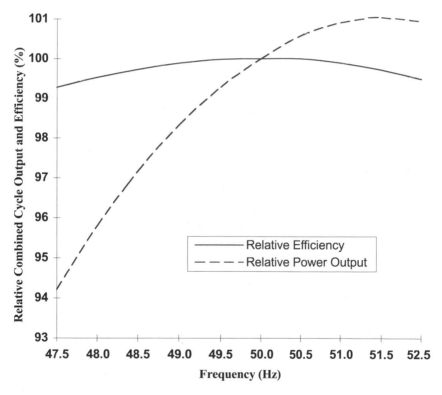

Figure 8-2 Effect of Frequency on Relative Combined-Cycle Output and Efficiency for Continuous Full Load Operation

Power factor: Plant power factors are dictated by the grid and influence the maximum generator capability and the generator efficiency, which in turn affects the output at the generator terminals. A normal power factor is in the range of 0.85 to 0.95. At the nominal point of the generator, a change in power factor from 0.8 to 1 at full load would improve the generator efficiency by 0.3 to 0.4%. For lower loads of the generator the difference tends asymptotically towards zero.

Process Energy

Plants with process extractions are often highly customized and general off-design behavior cannot be given.

Process energy has a major effect on plant performance, making a plant-specific curve necessary.

Fuel Type and Quality

The main off-design influence of the fuel on cycle performance occurs when a back-up fuel is fired (e.g., oil in place of natural gas). The reason for this lies in the fuel composition and possibly the need for water or steam injection to meet local emission requirements.

Variations in the composition for the same type of fuel also influence plant performance because a different fuel composition gives a different chemical composition after combustion.

Fuel components will also determine the lower heating value (LHV) of the fuel. The gas turbine fuel flow is defined as the heat input to the gas turbine divided by the LHV. If the LHV decreases, the fuel-mass flow increases to provide the same heat input to the gas turbine. This in turn results in increased flow through the gas turbine which has a positive impact on output.

However, if the chemical impact of the combustion products drag the performance in the opposite direction to the LHV influence, the total influence could be different.

Part-Load Behavior

In combined-cycles without supplementary firing, efficiency depends mainly on the gas turbine efficiency, the load of the gas turbine exhaust-gas temperature and plant size. So far, all of the corrections discussed have been related to a plant performance with the gas turbines running at full load with nominal firing temperature.

Figure 8-3 shows the part-load efficiency of a combined-cycle plant and the associated gas turbine, each relative to the 100% load case. At higher loads the part-load efficiency is good but this drops off more quickly below about 50% for two reasons.

First, the gas turbine used is equipped with three rows of compressor-variable inlet guide vanes, giving excellent part-load efficiency down to approximately 60% load. This because a high exhaust

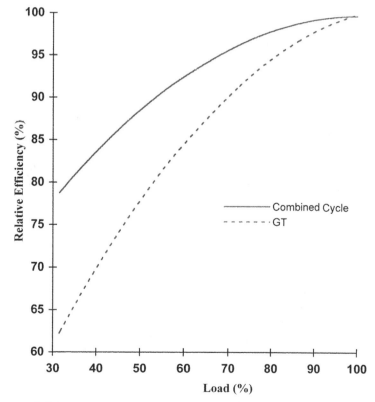

Figure 8-3 Part Load Efficiency of Gas Turbine and Combined-Cycle

gas temperature can be maintained as the mass flow is reduced. Below that level, the inlet temperature must be further reduced.

Second, the steam turbine is calculated with sliding pressure mode down to 50% load, also providing good utilization of the exhaust gas in this range. Below that point, the live-steam pressure is held constant by means of the steam turbine inlet valves, resulting in throttling losses.

For full-load operation, the gas turbine accounts for two-thirds of the power output and the steam turbine for one third. Figure 8-4 shows how the ratio of steam turbine to gas turbine power output (PST/PGT) shifts towards more steam turbine output at part loads. At 20% combined-cycle load, this ratio is actually reversed, with the steam turbine contributing two-thirds of the power output and the gas turbine, one-third.

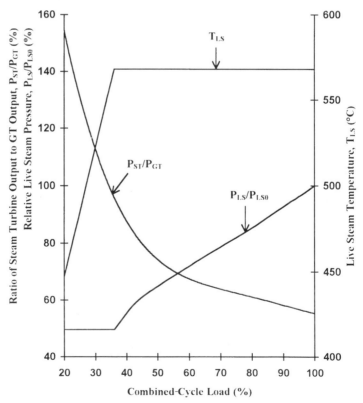

Figure 8-4 Ratio of Steam Turbine and Gas Turbine Output and Live Steam Data of a Combined-Cycle at Part Load

Additionally, the live-steam temperature and relative live-steam pressure of the water/steam cycle are shown. The live-steam temperature is kept constant by means of attemperation at the outlet of the HRSG superheater. The reason for the flat temperature profile lies in the variable inlet-guide vane control, which allows the gas turbine to operate at a lower flow with the nominal TIT, giving a higher exhaust gas temperature due to the lower pressure ratio of the unit. The live-steam pressure drops down to 50% of the full load live-steam pressure and is then controlled by the steam turbine valves.

There are several-site specific possibilities for further part-load efficiency improvements, such as:

- air preheating for sites with cold ambient air temperatures
- several gas turbines in the plant configuration

Figure 8-5 shows an example in which air is preheated using low-pressure steam from an HRSG allowing a higher relative gas turbine load. For gas turbines with VIGVs, there is practically no gain at

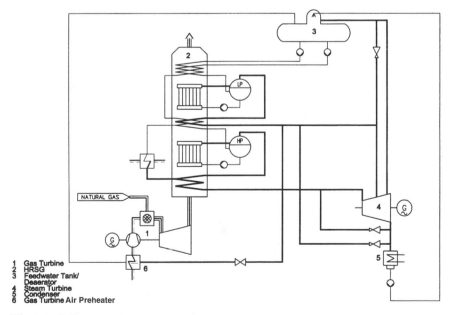

1 Gas Turbine
2 HRSG
3 Feedwater Tank/
  Deaerator
4 Steam Turbine
5 Condenser
6 Gas Turbine Air Preheater

Figure 8-5 Flow Diagram of a Dual Pressure Cycle with Gas Turbine Inlet Air Preheating

part-loads in incorporating this system. It could be an interesting alternative for older gas turbines that have fewer or no VIGVs, however, as it would allow the plant to operate at practically the same part-load efficiencies as for plants with VIGVs. Normally, the air can only be heated to approximately 40 to 50°C (104 to 122°F) without exceeding the limit imposed by compressor surge and the last-stage blade temperature limitation, meaning the efficiency gain drops off at hot ambient temperatures. Sometimes the system is utilized for modern gas turbines, not to achieve better part-load efficiency, but to maintain better emissions at lower loads.

A combined-cycle plant with several gas turbines is operated differently at part-load. For a plant with four gas turbines and one steam turbine, the overall plant load is reduced as follows:

- down to 75%, there is a parallel reduction in load on all four gas turbines
- at 75% one gas turbine is shut down
- down to 50%, there is a parallel reduction in load on the three remaining gas turbines
- at 50%, a second gas turbine is shut down
- etc.

With this mode of operation, the efficiency at 75%, 50%, and 25% load is lower than at full load. If, however, four independent single-shaft combined-cycle blocks are selected, the part-load efficiencies are as shown in Figure 8-6. In that case, the full load efficiency would be achieved at the points 100%, 75%, 50%, and 25 % load because at these points the individual steam turbines are also running at full load.

Combined-Cycle Testing Procedures

When a power plant is sold, plant performance guarantees are given. These guarantees will apply to a set of performance parameters and ambient conditions that cannot normally be recreated for the performance test. This means that, in order to demonstrate that the guarantees have been met, plant performance must be measured under the conditions that exist at the site at the time and the results must be cor-

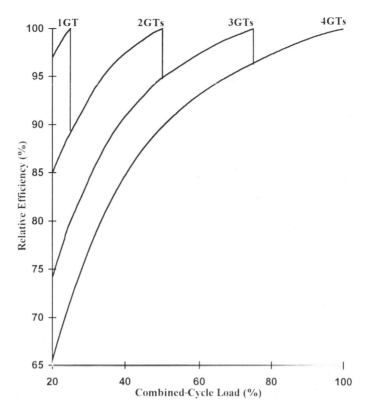

Figure 8-6 Part Load Efficiency of Combined-Cycle Plant with Four Single Shaft Blocks

rected with a series of corrections for the parameters described earlier. This corrected value is then compared to the guaranteed values.

In a combined-cycle plant the gas turbine, HRSG, and steam turbine all interact with one another. If the contract for the plant goes to a single general contractor on a turnkey basis, the power output and efficiency of the plant as a whole can be guaranteed. Similarly, for combined-cycle plants, it is easier to measure the values for overall plant performance than those for each major component individually. The amount of waste heat supplied to the HRSG by the gas turbine, in particular, cannot be measured accurately. When overall values are guaranteed, the fuel flow, electrical output, and ambient conditions of the power plant must be measured. These are quantities that can be determined with relative exactness.

Thereafter, a correction factor is determined for each parameter, quantifying its influence on actual performance due to the fact that each parameter is not at the design/guarantee value. These correction factors are multiplied together to give values which can be directly compared to the guarantees.

The power output of the gas turbine and the steam turbine are often corrected separately.

- For the gas turbine, the usual correction curves are used to take into account the effects produced by air temperature, air pressure, rotational speed, etc
- The power output measured for the steam turbine is corrected using curves that show the indirect effects of air temperature, air pressure, and gas turbine speed on the steam process and the direct effect of the cooling water temperature

To calculate these curves, it is best to use a computer model that simulates the steam process as a whole (as described previously in this chapter). Changes in ambient air data produce changes in the gas turbine exhaust data and these affect the power output of the steam turbine.

The advantage of this procedure is that it can, with certain modifications, be used even if the gas turbine is put into operation at a somewhat earlier date than the steam turbine. The method is, however, rather complicated and requires quite a few corrections to cover the interactions between the gas turbine(s) and steam turbine. Therefore–especially for single-shaft combined-cycles and for plants without phased construction–there is a trend towards overall combined-cycle corrections.

At times, performance predictions after a certain number of operating hours or years of operation are required. After commissioning, the plant is normally said to be in "new and clean condition". For operating hours beyond this point, degradation is considered in predicting the performance. The degradation is mainly caused by the gas turbine and partially recuperated in the steam process. Typical values are given in chapter 2.

For steam turbine power plants and for gas turbines and combined-cycle power plants, the methods used for corrections are described in international standards (e.g., ASME and ISO).

ENVIRONMENTAL CONSIDERATIONS

CHAPTER 9

9

Environmental Considerations

The impact any power plant has upon its environment must be minimized as much as possible. Legislation in different countries establishes rules and laws that have to be fulfilled. Quite often emission limits are based upon the best available emission-control technology. Exhaust emissions to the environment are mainly controlled in the gas turbine. Often, regions with less-stringent air emission requirements profit from the same combustion technology as areas with stringent requirements because the same hardware is used.

The following emissions from a power station directly affect the environment:

- combustion products (exhausts and ash)
- waste heat
- waste-water
- noise
- radioactivity and nuclear waste (nuclear power stations)

Exhausts can include the following components: H_2O, N_2, O_2, NO, NO_2, CO_2, CO, C_nH_n (unburned hydrocarbons, UHC), SO_2, SO_3, dust, fly ash, heavy metals, and chlorides.

The first three of these are harmless; the others can negatively affect the environment. Concentration levels of these substances in the exhaust gas depend on the fuel composition and the type of installation. However, the greater the efficiency of the installation, the greater the drop-off in the proportion of emissions per unit of electrical energy produced.

Because most combined-cycle plants burn natural gas, they produce low exhaust emissions. Their high efficiency results in low air emissions per MWh of electrical power produced and a low amount of waste heat. The high excess air ratios customarily found in gas turbines enable practically complete combustion to take place, which results in a very low concentration of unburned elements such as CO or UHC in the exhaust. Also due to the extremely low sulfur content in natural gas, SO_x (SO_2, SO_3) emissions are negligible. Therefore, a combined-cycle plant can be considered to be environmentally friendly and well suited for use in heavily populated areas.

For plants burning natural gas, the most relevant emissions in the exhaust are NO and NO_2. NO_x (NO and NO_2) emissions generate nitric acid (H_2NO_3) in the atmosphere which together with sulfuric and sulfurous acids (H_2SO_4, H_2SO_3), are factors responsible for acid rain. CO_2 is created by burning fossil fuels and is held responsible for global warming.

Reduction of NO_x Emissions

NO_x is produced in air in large quantities at very high temperature levels. Figure 9-1 shows NO_x concentrations in the air as a function of air temperature. The concentration shown is at equilibrium in air attained after an infinite time.

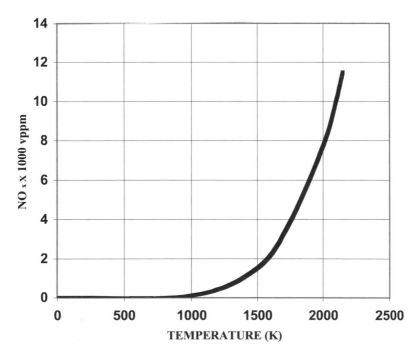

Figure 9-1 NO_x Equilibrium as a Function of Air Temperature

The situation in a gas turbine combustor is different, because combustion between fuel and air takes place and also because high-temperature residence time is fairly limited. The major factors affecting NO_x production in the combustor are:

- fuel to air ratio of the combustion
- combustion pressure
- air temperature into combustion chamber
- duration of the combustion

As can be seen from Figure 9-1, NO_x is formed only when temperatures are high; this is the case in the flame of the combustor. The temperature of this flame depends on the fuel to air ratio of the flame and the air temperature into the combustion chamber, as shown in Figure 9-2. It is highest in the case of stoichiometric combustion, (fuel to

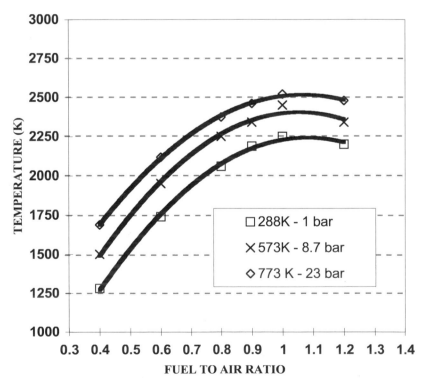

Figure 9-2 Flame Temperature as a Function of the Fuel to Air Ratio and Combustion Air Conditions

air ratio = 1). Figure 9-3 shows how concentrations of NO_x depend on the fuel to air ratio and the combustion air conditions.

It is evident that a peak is reached with a ratio of approximately 0.8. Above that level, the flame temperature is higher but there is less oxygen available to form NO_x, since most of it is used for the combustion. Below that level, NO_x decreases because of the abundance of excess air within the flame, lowering the flame temperature.

Very high fuel to air ratios are beneficial from the point of view of NO_x but are detrimental to combustion efficiency and cause the production of large amounts of CO and unburned hydrocarbons (UHC).

Conventional gas turbine combustors with a diffusion burner were designed to operate with a fuel to air ratio of approximately 1 at full load, ensuring good stable combustion over the entire load range. Obviously, NO_x emissions are high unless special precautions are

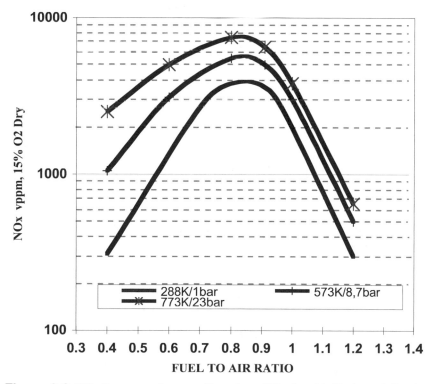

Figure 9-3 NO_x Concentration as a Function of Fuel to Air Ratio and Combustion Air Conditions

taken. Nowadays burners operate at a lower fuel to air ratio that results in lower NO_x emissions.

The simplest way to reduce NO_x concentrations in these diffusion burners is to cool the flame by injecting water or steam into it. Figure 9-4 shows the reduction factors for NO_x emissions that can be attained as a function of the amount of water or steam injected. The amount of water or steam injected is indicated by the coefficient Ω (the ratio between the flows of water or steam and fuel). At a ratio of $\Omega=1$ the typical reduction factor is approximately 6 with water and approximately 3 with steam. Water is more efficient than steam because evaporation takes place in the flame at low temperature, providing effective cooling.

With this wet method, it is possible to attain NO_x levels as low as 40 ppm (15% O_2 dry) in the exhaust gases from gas-fired gas turbines or a combined-cycle.

Steam or water injection is a simple way to reduce NO_x emissions but it does entail the following disadvantages:

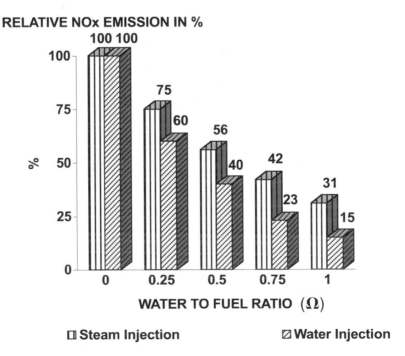

Figure 9-4 NO_x Reduction Factor as a Function of the Water or Steam to Fuel Ratio in Gas Turbines with Diffusion Combustion

- large amounts of demineralized water are required
- the efficiency of the combined-cycle plant is lower, particularly if water injection is applied

The fact that these methods can increase the plant output (especially with water injection) can be of interest in plants with a low number of operating hours per year. In general, however, this does not compensate for the loss in efficiency and the high water consumption.

Table 9-1 shows how steam and water injection affect the output and efficiency of a combined-cycle plant as a function of the water or steam to fuel ratio. With a ratio of $\Omega=1$, the following changes in output and efficiency, compared with a dry cycle without injection, may be considered as typical:

Table 9-1 Output and Efficiency of a Combined-Cycle Plant with Cold Water or Steam Injection Compared to the Same Plant without Injection

	Change in Efficiency, %	Change in Output, %
Water injection, $\Omega=1$	-4.0%	+9.0%
Steam injection, $\Omega=1$	-1.5%	+4.8%

The disadvantages inherent in steam or water injection have motivated all gas turbine manufacturers to develop combustors which attain low NO_x levels with dry combustion (i.e., without injecting steam or water).

The principle behind keeping NO_x levels low is to always dilute the fuel with as much combustion air as possible to maintain a low flame temperature, and to keep the resident time in the hot combustion zone short. With this "dry low-NO_x method" and firing with natural gas, NO_x levels of 25 ppm (15% O_2 dry) are achieved with modern gas turbines of high firing temperatures and efficiencies.

Local regulations in some parts of the United States and in Japan require NO_x emissions well below 25 ppm (15% O_2 dry). In these cases, it is generally necessary to install a reduction system in the HRSG. Known as Selective Catalytic Reduction (SCR), these systems inject ammonia (NH_3) into the exhaust gases upstream of a catalyst and can thereby remove up to approximately 85% of the NO_x in the exhaust gases leaving the gas turbine. The chemical reactions involved are as follows:

$$4 \text{ NO} + 4 \text{ NH}_3 + O_2 = 4 \text{ N}_2 + 6 \text{ H}_2O \qquad \text{Eq. 9-1}$$
$$6 \text{ NO}_2 + 8 \text{ NH}_{3 +} + = 7 \text{ N}_2 + 12 \text{ H}_2O \qquad \text{Eq. 9-2}$$

Technically these are well-proven systems, but they entail the following disadvantages:

- investment costs are high; the HRSG is 10-30% more expensive
- the use of ammonia is necessary, with a fraction of the ammonia passing through the SCR (ammonia slip)
- power output and efficiency of the power plant are reduced by approximately 0.3% due to the increased back-pressure of the gas turbine
- when firing oil in the gas turbine, sulfur in the fuel reacts to form ammoniabisulphate, which precipitates in the cold end of the HRSG and further increases back-pressure to the gas turbine. Power output and efficiency of the plant is further reduced
- the HRSG requires periodical cleaning, the waste must be disposed of
- the catalyst must be installed in the evaporator section of the HRSG since the reaction takes place only in a temperature window of 300 to 400°C (572 to 752°F)
- replacement costs are high

Figure 9-5 shows a typical SCR system installed in an HRSG.

NO_x levels of less than 5 ppm (15% O_2 dry) can be achieved by applying a SCR system in conjunction with a gas turbine equipped with dry low-NO_x burners.

SO$_x$ Emissions

Concentrations of SO_2 and SO_3 produced depend only on the quality of the fuel. The majority of gas turbines use clean natural gas fuel, therefore the SO_x emissions are negligible.

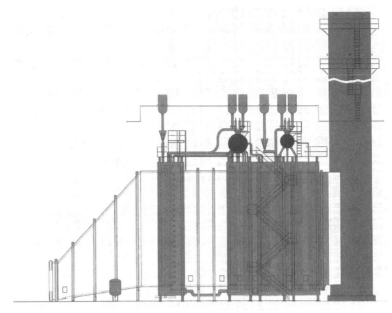

Figure 9-5 Heat Recovery Steam Generator with Selective Catalytic Reduction

Liquid fuels contain between 0.05% sulfur for sulfur-reduced No.2 oil and 2% sulfur for heavy oil. If the resulting SO_x emissions are not acceptable the most economic way to reduce the emissions is to directly treat the sulfur content in the fuel either by removing sulfur or by blending the fuel with a fuel of lower sulfur content.

CO_2 Emissions

Each fossil-fired power plant produces CO_2, which is held responsible for global warming. However, a modern combined-cycle plant burning natural gas produces approximately 40% of the CO_2 per MWh of electricity of a conventional coal fired power plant for two reasons:

- higher efficiency
- use of natural gas, which is mainly methane (CH_4) as opposed to coal (C)

During the 1990s, with deregulation in the power generation market in the U.K. modern, high-efficiency, gas-fired combined-cycle plants replaced many old coal-fired power stations. This replacement meant that CO_2 production per MWh of electricity of these new power plants dropped to a third of the value for the coal fired plant. This example shows that deregulation can have a strong positive ecological effect.

In the U.S., a very significant portion of the installed coal-fired plants has been in service for more than 30 years. Unless coal is available at low cost a similar effect to that in the U.K.will take place. This is shifting capacity to gas-fired combined-cycle plants as deregulation takes place and to gas turbine peaking plants for peak power in summer.

Waste Heat Rejection

Another environmental concern is the waste heat that every thermal power station releases to the environment. Here too, the high efficiency of the combined-cycle plant is an advantage: from any given amount of primary energy, a greater amount of electricity or useful output is produced, which reduces the amount of waste heat to the environment.

In addition to the quantity of waste heat, however, the form in which the heat is given to the environment is also important. The effect is less if the power plant heats air instead of giving off its waste heat to a river or the sea. Conventional steam power plants often dissipate the waste heat to water for efficiency reasons. The most economic solution for combined-cycle plants is frequently to dissipate the heat to the air through a wet cooling tower. Direct air-cooling is also possible, but results in a reduction in output and efficiency together with increased costs due to the air-cooled condenser.

A combined-cycle plant needs only half the cooling water of a conventional steam plant of the same output and a third of what is required for a nuclear power station.

A gas turbine usually requires practically no external cooling except for the lube oil and generator, which has contributed greatly to its widespread acceptance in countries where water is scarce. Due to the fact that the steam portion of the combined-cycle plant accounts only for a third of the output, an air-cooled condenser is an economic solution for dry areas underlining the wide range of possible applications. Table 9-2 shows the amount of waste heat that must be dissipated for differnet types of plant with 1,000 MW electrical output. All steam cycles are cooled by river or seawater.

Table 9-2 Comparison of the Heat to be Dissipated for Differnet Types of 1,000 MW Station

Heat sink	Gas turbine	Combined-cycle	Steam turbine	Nuclear
air/stack	1,500-2,000 MW	130-180 MW	70-100 MW	0 MW
water	0 MW	550-700 MW	1,100-1,400 MW	1,800-2,200 MW

Noise Emissions

An environmental issue that is considered during the design and construction of a combined-cycle power plant is noise. This issue can be solved using the acoustic insulation and silencing equipment available today.

A distinction is made between *near field noise* and *far field noise*. *Near field noise* refers to the noise levels at the machinery. *Far field noise* often refers to the plant boundary and indicates the noise emitted to the neighborhood. Main sources for this noise are the gas turbine inlet, the gas turbine exhaust (or HRSG), the stack and cooling tower, or an air-cooled condenser. Steam bypass operation during start-up and shut down is an additional source of noise.

Conclusion

Due to its low emission levels, low cooling requirements, and noise levels that can meet stringent requirements, combined-cycle plants are considered to be environmentally friendly and are well suited for decentralized power generation in urban areas.

Developmental Trends

CHAPTER 10

10

Developmental Trends

There are five main trends in the continued development of combined-cycle power plants.

- increased gas-turbine firing temperatures to achieve higher exhaust gas temperatures and higher efficiencies for the water/steam cycle
- new gas turbine concepts that will achieve a higher efficiency and greater output for the plant
- increased gas turbine power output to reach lower specific power plant costs
- higher live-steam parameters in the steam/water cycle to improve efficiency
- lower emissions, especially NO_x, to reduce environmental impacts

These trends are continuations of developments that have led to the breakthrough of combined-cycle plants in the past.

Firing Temperatures

The positive effect of high gas-turbine inlet temperatures on the efficiency of the combined-cycle and the consequent reduction in the fuel cost part in the cost of electricity is pushing gas-turbine inlet temperatures further up.

It seems reasonable to look for further improvements from even higher gas turbine inlet temperatures that have become possible through the development of new materials and improved cooling systems. Research projects in this area are mainly concentrated on improved air cooling of the hot-gas path of the gas turbine. Alternative cooling technologies employing steam cooling, for example, have been announced by gas turbine manufacturers.

The use of ceramic materials in gas turbines, mainly for blading, still appears far from ready for commercial operation in large gas turbines for power generation. Ceramics are currently used for certain parts of the hot-gas path. Ceramic-based thermal barrier coatings allow hot-gas temperatures to be raised while keeping metal temperatures at the same level.

Sequential Combustion

A gas turbine with a sequential combustion chamber comprises a compressor, an HP combustion chamber, an HP turbine, an LP combustion chamber, and an LP turbine resulting in a higher compression ratio than in a gas turbine with single combustion. Due to the higher pressure ratio and sequential combustion, the mean temperature of the heat supplied is higher. This raises the efficiency of the combined-cycle plant. An important benefit is the high exhaust-gas temperature of the gas turbine that provides the basis for a highly efficient steam-turbine cycle.

The two combustion chambers, in combination with the inlet guide vanes to the compressor, allow the gas turbine exhaust temperature to be kept constant at part-load operation. This has two major benefits.

First, it results in excellent part-load efficiency of the whole combined-cycle plant which is important because these plants are often operated in cycling duties. In power generation markets, purchase prices for electricity in low tariff hours can drop below the cost of power generation. This forces a power plant to shut down (e.g., at night) and restart when tariffs are higher (e.g., the next morning). This procedure results in "start penalties" for equipment warm-up and additional fuel costs. A high part-load efficiency allows the plant manager to compete at times of low dispatch since the variable cost remains low and the start penalty can be avoided.

Second, due to the constant exhaust temperature, the steam temperature for the steam/water cycle also remains constant, which results in limited cycling stresses for the steam turbine and HRSG when changing from part-load to full-load or vice versa.

Gas turbines with sequential combustion chambers were first introduced in the late 1940s and reintroduced in the early 1990s. Several gas turbines of the latter design are now in commercial operation.

Closed Steam Cooling of Stationary Parts

The majority of gas turbines presently in operation have air-cooled blades and combustion chambers.

Cooling of the blading allows turbine inlet temperatures to be increased beyond the allowable metal temperatures. If air-cooling is employed, the cooling air first cools the blade internally and then exits onto the blade surface, providing a cooling film that separates the hot gas from the blade surface. However, the mixing of the two flows leads to a disturbance of the flow pattern and to a cooling of the hot gas, which reduces the efficiency. An alternative is to use steam as a cooling medium. Closed-steam cooling takes advantage of the fact that steam has a higher heat capacity than air and doesn't deteriorate the expansion efficiency. Additionally, more air is available for the combustion process, which results in a lower flame temperature and lower NO_x levels. Additional output is achieved by the increased mass flow through the turbine.

Steam for cooling is available from the steam part of the water steam cycle. It can be taken from the cold reheat line exiting from the steam turbine.

An increase in plant efficiency can be achieved with closed-steam cooling. However, as the gas turbine and steam turbine cycle get more interconnected, there are corresponding disadvantages, such as restricting or even preventing operation of the plant as a simple-cycle gas turbine.

The first plant based on this technology is now in the testing phase and is due for commercial operation before the end of the decade.

Steam Cooling of Stationary and Rotating Parts

The improvement potential is even larger for steam cooling if the rotating parts are steam-cooled as well.

A combined-cycle plant with a fully steam-cooled gas turbine is expected to offer a higher efficiency and offers a larger specific power output per unit of inlet air flow. This could result in lower specific costs although maintenance costs are expected to be higher due to the higher blading costs. The following points are challenges for this type of gas turbine:

- cooling of the thin front and rear ends of the blades
- steam purity requirements for the cooling steam
- leakage tightness of the steam system (steam must be supplied to the gas turbine rotor and returned from there)

This type of plant is attractive from an economic point of view only if these challenges are solved satisfactorily and the plant achieves the corresponding reliability.

Larger Compressors

Parallel to these developments, improvements are also being made to the compressor. The advantages offered by the higher gas temperatures can only be fully exploited if the pressure ratio of the machine is increased to an appropriate level. Increasing the airflow through the compressor also attains high unit ratings. With modern blading, compressors are able to handle volume flows that seemed utopic just a few years ago. With modern compressor designs it is already possible to increase pressure ratios and reduce the number of compressor stages.

Higher Live-Steam Parameters

Early combined-cycle plants had water/steam cycles with low steam parameters compared to conventional steam power plants. Live-steam pressures in a range of 50 to 80 bar (710 to 1,150 psig) and live-steam temperatures of 450 to 500°C (842 to 932°F) were standard values. Combined-cycle power plants with single-pressure steam generation even used pressures in a range below 50 bar (710 psig).

Modern, large combined-cycle power plants for power generation apply a live-steam pressure of 100 to 160 bar, (1,430 to 2,310 psig) and a live-steam temperature of 520 to 565°C (968 to 1,049°F). For steam turbines beyond approximately 100 MW output, reheat machines are now standard compared to the previous non-reheat units.

This was made possible by the sequential firing technology or higher firing temperature of the gas turbines which, as a result, provided higher exhaust temperatures and therefore higher inlet temperatures to the HRSG. Higher gas-turbine exhaust temperatures yield a higher optimal live-steam pressure based on the net present value (NPV) of the plant. In the near future it can be expected that these trends will continue. Large plants for power generation will include higher live-steam pressures, increased live-steam temperatures, and reheat for the steam cycle.

The following points must be considered when the live-steam parameters are increased:

- higher live-steam temperatures require more expensive alloys in the HRSG, steam piping, and steam turbine. The gain in output must therefore justify this additional investment
- higher live-steam pressures cause wall thickness to increase which, in general, reduces thermal flexibility and increases cost. Once-through HRSGs will be installed more frequently to avoid the negative impact on thermal flexibility of the higher live-steam pressures
- high live-steam pressure in combination with a reheat steam turbine reduces the live-steam volume, which may cause a reduction in the efficiency of the high-pressure steam turbine due to short blading

The last effect would lead to a live-steam pressure that is too far away from the thermodynamic optimum. To avoid this either several gas turbines and HRSGs are combined to one steam turbine or the high-pressure steam turbine runs at higher speed, which reduces the rotor diameter and increases blade length. A side-benefit is the smaller casing dimensions, which improve thermal flexibility. Figure 10-1 shows a geared high-pressure turbine.

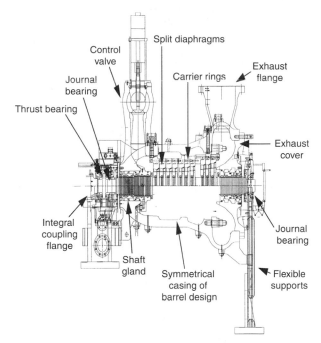

Figure 10-1 Geared High Pressure Turbine

As the exhaust-gas temperatures of the gas turbine rise further, the optimal cycle selection is affected.

The triple-pressure reheat cycle is often the optimal steam cycle for a gas turbine with an exhaust temperature of approximately 600°C (1,112°F). The optimal cycle will move towards a single-pressure re-heat cycle for gas turbines with an exhaust temperature of 750-800°C (1,382-1,472°F).

Lower NO$_x$ Emissions

Theoretically, there are two ways that lower NO$_x$ emissions can be achieved:

- oxygen-lean combustion
- oxygen-rich combustion

Figure 9-3 showed the dependence of NO$_x$ concentrations on the fuel to air ratio. Despite a high flame temperature, only a small amount of NO_x can form in oxygen-lean combustion because there is scarcely any oxygen available to produce NO_x. However, in order to attain complete combustion, there must be a second, follow-up combustion stage here in which almost no NO_x is formed, due to the lower temperature. This approach is being used in modern steam generators and is referred to as "staged combustion".

For gas turbines with dry low NO_x burners, a different–and more effective–procedure is applied because of the high overall excess air ratio (2.5 to 3.5). It is called *combustion with excess air* and is the same principle as the injection of water or steam, where a large amount of excess air effectively cools the flame. The procedure is subject to limits, due to consideration of flame stability. With excess air ratios of approximately three, the combustion becomes very poor and the flame is completely extinguished.

This risk does not exist while the gas turbine is at full load, because there is not enough air available for the burner to significantly exceed an excess air ratio of two. The remaining air is needed to cool the hot parts and the turbine blading. It is, however, a problem at part-loads.

For combustion to actually take place at the desired excess air ratio, the air and fuel must be mixed homogeneously with each other. Figure 10-2 shows a typical low-NO_x burner. The air and natural gas are premixed in the burner and are burned downstream of the burner. A vortex breakdown structure holds the flame in the free space. Characteristics of this burner include:

- no flashback due to the vortex-breakdown structure
- low fuel to-air ratio (and, therefore, low NOx)
- for liquid fuel, water injection is used to limit NO_x emissions

With this combustion technology, there are modern gas turbines with high firing temperatures in operation with NO_x emissions in the range of 25 ppm NO_x at 15% O_2.

Further development of the low-NO_x burners is proceeding towards even lower emission levels. Due to investments in combustion

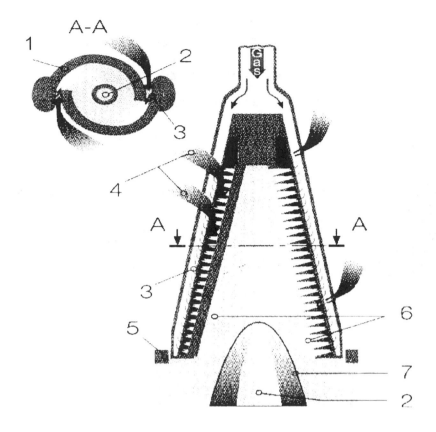

1 Vortex Generator	5 Burner Exit Level
2 Vortex Core	6 Mixing of Gas and Combustion Air
3 Gas Injection	7 Flame Front
4 Combustion Air	

Figure 10-2 Cross Section of a Low NO_x Burner

research, lower NO_x levels are attainable and potential for further improvement is assumed. Figure 10-3 shows a cross-section of such a low-NO_x combustion chamber.

With these improvements, combined-cycle plants with a guaranteed net efficiency of 58% are already in commercial operation. Gas turbines with unit power capacities of more than 300 MW for 50 Hz applications and 200 MW for 60 Hz applications mean lower costs, so

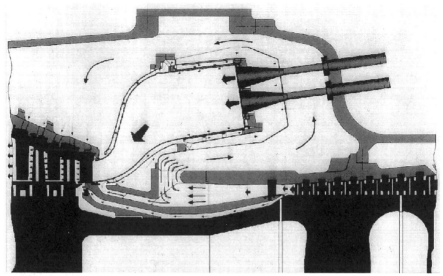

Figure 10-3 Cross Section of a Gas Turbine Combustion Chamber

that combined-cycle plants will become more competitive for large power plants. It can be expected that just after the turn of the century, gas fired combined-cycle plants can be offered with net efficiencies of 60%, based on the lower heating value of the fuel.

SOME TYPICAL COMBINED-CYCLE PLANTS

CHAPTER 11

11

Some Typical Combined-Cycle Plants

The following examples show how the technologies previously described are applied in the global electric power industry.

The 1300-MW Combined-Cycle Plant in Lumut, Malaysia

The Lumut combined-cycle power plant is owned by Segari Energy Ventures Sdn. Bhd. (SEV), a subsidiary company of Malakoff Berhad. The plant is located on an 80-acre site on old tin mining land at Mukim Pengkalan Baru, District of Manjung in Lumut, Malaysia. Lumut combined-cycle power plant is designed for pure power generation, with a net output of 1,303 MW, making it Malaysia's biggest gas fired power station.

To safeguard the environment the power plant was designed with the latest technology to reduce emissions and effluents to levels well below the requirements of the local Department of Environment.

The Lumut power plant is made up of two identical 651.7 MW blocks, each consisting of three 143 MW ABB GT13E2 gas turbines, three ABB CE natural circulation HRSGs, and one ABB 231 MW double-casing condensing steam turbine generator unit (Fig. 11-1).

The plant has a typical dual-pressure cycle, where HP (67 bar/957 psig) / 508°C/946°F) and LP (5.9 bar/71 psig) / saturated) steam is generated in the HRSGs before being fed to the common steam turbine. In the HRSGs the hot exhaust gases of the gas turbines 544°C (1,011°F) and 485 kg/s (3,850,000 lb/h) are cooled to a temperature below 100°C (212°F) before leaving through the stacks (Fig. 11-2).

A direct seawater cooling system cools the entire power plant. The seawater is taken from a point 1,000 meters from the seashore and is led to the cooling water intake basin through four 2.8-meter diameter steel pipes. Four 300-meter steel pipes return the cooling water to the sea with a cooling water temperature rise below 7 K (12.6 R).

Figure 11-1 View of Lumut Combined-Cycle Power Plant

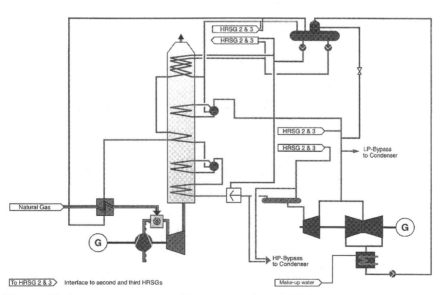

Figure 11-2 Process Diagram of Lumut Combined-Cycle Power Plant

In Table 11-1, the performance of the Lumut combined-cycle power plant is summarized. As shown, more than 52% of the fuel energy is converted to electrical power, with NO_x emissions below 25 vppm (15% O_2 dry) when firing natural gas. Further, the noise level at the plant boundary does not exceed 65 dBA.

Table 11-1 Main Technical Data of the Lumut Combined-Cycle Power Plant

Plant configuration:		2 x KA13E2-3
Number of blocks		2
Number of Gas Turbines/block		3
Main fuel		Natural gas
Ambient conditions:		
Ambient temperature	°C / °F	32 / 89.6
Ambient pressure	mbar	1,013
Relative humidity	%	80
Cooling water temperature	°C / °F	32 / 89.6
Performance data per block:		
Total Fuel Input to the Gas turbines	MW	1,261.5
Gas Turbine Power Output	MW	429.3
Steam Turbine Power Output	MW	231.2
Gross Output	MW	660.5
Gross efficiency (LHV)	%	52.4
Auxiliary consumption and losses	MW	8.8
Net Power Output	MW	651.7
Net Efficiency (LHV)	%	51.7
Heat rate (LHV)	kJ/kWh	6,969
Heat rate (LHV)	Btu/kWh	6,605
Process heat	MJ/s	0

The electrical energy produced in the gas-and-steam turbine generators is transformed in the main transformers to the required grid voltage and is exported via the power plant switchyard to TNB's National Grid via 275 kV and 500 kV transmission lines (Fig. 11-3).

The Lumut power plant is arranged in a traditional manner with separate gas turbine and steam turbine buildings. The HRSGs are of outdoor type with a small roof protecting equipment against rain. The balance of the plant equipment is decentrally situated surrounding the power blocks giving logical flow directions.

Overall project development, implementation and key events are shown in Table 11-2. The table gives an example of the phases in a typical IPP project with a comfortable delivery time.

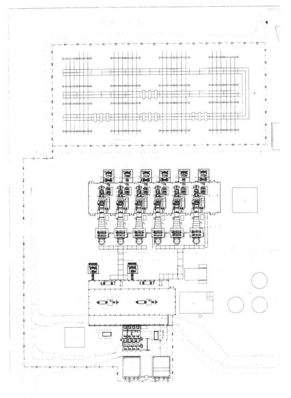

Figure 11-3
General Arrangement of the Lumut Combined-Cycle Power Plant

Table 11-2 Overall Project Development of the Lumut Combined-Cycle Power Plant

Key Events:	
Obtaining IPP license	15.07.1993
Signing of Fuel Supply Agreement	17.07.1993
Signing of Power Purchase Agreement	16.10.1993
Signing of EPC contract	02.12.1993
Signing of O&M contract	03.12.1993
Financial Closing	12.04.1994
First ignition of first gas turbine, block 1	30.01.1996
First synchronization of first gas turbine, block 1	06.02.1996
First steam to steam turbine, block 1	28.03.1996
First synchronization of block 1	06.04.1996
Commercial operation of block 1	28.05.1996
First ignition of first gas turbine, block 2	19.08.1996
First steam to second steam turbine, block 2	09.10.1996
First synchronization of block 2	12.10.1996
Commercial operation of block 2	10.01.1997

EPC = Engineering Procurement and Construction,
IPP = Independent Power Producer
O&M = Operation and Maintenance

The Diemen Combined-Cycle Cogeneration Plant, Netherlands

N.V. Energiproduktiebedrijf UNA owns the Diemen 33 combined-cycle power plant which went into operation in the autumn of 1995. The plant is located in the Netherlands close to the town of Muiden, Southeast of Amsterdam and consists of one combined-cycle block with one ABB GT13E2, one HRSG and a triple-pressure reheat steam turbine providing electrical power and heating to the Southeastern part of Amsterdam.

In designing the plant, a high degree of flexibility was required, allowing the plant to operate in purely condensing mode with an electrical output of 249 MW or in a combined-power and district-heating mode with 218 MW electrical output and 180 MJ/s district heating production (Fig. 11-4).

A high-efficiency triple-pressure reheat cycle was chosen, in which the HP live-steam (89 bar (1,276 psig) / 505°C (941°F) / 49 kg/s (389,000 lb/h) is fed to the stand-alone HP steam turbine, where it expands and mixes with the IP steam before being reheated in the HRSG (24.5 bar (340 psig) / 505°C (941°F) / 61 kg/s (484,000 lb/h) and fed back to the steam turbine. In the IP steam turbine the steam is expanded to the LP level, where it mixes with the LP steam (4.6 bar (52 psig) / saturated / 9.3 kg/s (74,000 lb/h) before undergoing final expansion (Fig. 11-5).

To optimize the district-heating mode (winter operation, mainly), three stages of district heating were provided, where the first stage is supplied with a water extraction from the HRSG and the second and third stages with steam extractions from the steam turbine. For the condensing mode (summer operation, mainly), a double-flow LP steam turbine was chosen to take full advantage of the cold cooling water conditions in the direct cooling system.

Figure 11-4 View of Diemen Combined-Cycle Cogeneration Plant

Table 11-3 Main Technical Data of the Diemen Combined-Cycle Cogeneration Plant

Plant configuration:		KA13E2-1	
Number of blocks		1	
Number of Gas Turbines / block		1	
Main fuel		Natural gas	
Ambient conditions:			
Ambient temperature	°C / °F	15 / 59	
Ambient pressure	mbar	1013	
Relative humidity	%	50	
Cooling water temperature	°C / °F	15 / 59	
		"Summer mode":	**"Winter mode":**
Performance data			
Total Fuel Input to the Gas Turbines	MW	455.5	455.5
Gas Turbine Power Output	MW	161.6	161.6
Steam Turbine Power Output	MW	91.1	59.8
Gross Output	MW	252.7	221.4
Gross efficiency (LHV)	%	55.5	48.6
Auxiliary consumption and losses	MW	3.5	3.1
Net Power Output	MW	249.2	218.3
Net Efficiency (LHV)	%	54.7	47.9
Heat rate (LHV)	kJ/kWh	6580	7512
Heat rate (LHV)	Btu/kWh	6237	7120
Process heat	MJ/s	0	179.6
District heating return temperature	°C / °F	N.A.	65 / 149
District heating forwarding temperature	°C / °F	N.A.	105 / 211
Fuel utilization (LHV)	%	N.A.	87.4

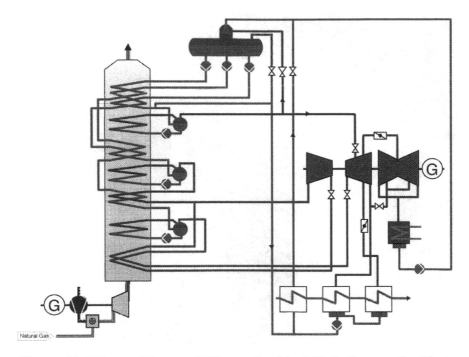

Figure 11-5 Process Diagram of Diemen Combined-Cycle Cogeneration Plant

The single annular burners of the 13E2 gas turbine restrict NO_x emissions on gaseous fuels to below 25 vppm (15% O_2 dry). To further reduce the NO_x emissions at part-load, LP steam is supplied to a heat exchanger at the gas turbine air intake allowing the gas turbine to be operated at nominal TIT for an even wider load range. For temperatures between -7 and +7°C, the heat exchanger is operated to avoid icing at the compressor inlet, which again gives a better efficiency than conventional anti-icing systems where hot air is extracted from the compressor and fed back to the gas turbine air intake.

To give maximum weather protection to the gas turbine, HRSG and steam turbine they are arranged indoors. The gas turbine and HRSG are floor-mounted and the steam turbine generator unit is mounted on a table with the condenser and district heaters situated beneath it. Plant internal electrical consumers are fed from a common electrical room fed from the auxiliary transformers, connected to the common steam turbine and gas turbine three-winding step-up transformer (Fig. 11-6).

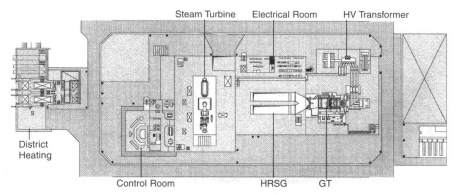

Figure 11-6 General Arrangement of Diemen Combined-Cycle Cogeneration Plant

The 347-MW Combined-Cycle Plant in King's Lynn, United Kingdom

The King's Lynn combined-cycle power plant was ordered from Siemens plc in 1995 by Anglian Power Generators, Ltd., a subsidiary of Eastern Generation, Ltd. The plant is located just south of King's Lynn, Norfolk, about 120 km north of London. The site was selected due to its proximity to the British Gas Transco pipeline and the 132 kV grid. The power plant is designed for pure power generation with a gross output of 347 MW. The main fuel is natural gas with distillate oil as a back up.

The King's Lynn power plant is made up of one single-shaft turbine generator unit, with one Siemens V94.3 gas turbine, one vertical-forced circulation drum-type HRSG, and one Siemens two-casing reheat-condensing steam-turbine unit. The generator is hydrogen cooled. Although there is a water source nearby, restrictions on its use prevented water-cooling so an air-cooled condenser was installed (Fig. 11-7).

The cycle is a triple-pressure reheat cycle, with HP steam at 101 bar (1,450 psig) / 519°C (966°F), reheat steam at 26 bar (362 psig) / 518°C, (964°F) and saturated LP steam at 5.0 bar (58 psig). No feedwater tank is installed but a partial-flow deaerator is provided for start-up. Fuel preheating using an extraction of hot water from the HRSG increases the cycle efficiency (Fig. 11-8).

Table 11-4 summarizes the performance data of the King's Lynn combined-cycle power plant.

Figure 11-7 King's Lynn Power Station GUD 1S.94.3 (status 9/96)

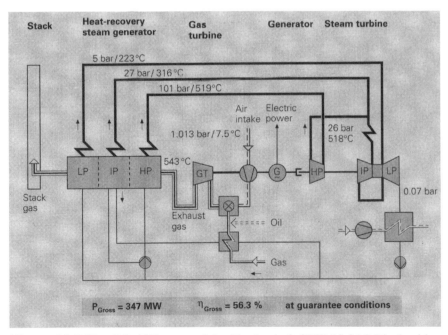

Figure 11-8 Simplified Diagram of the Single-shaft GUD 1S.94.3 Block with a Triple-pressure Reheat Steam Cycle and Air-cooled Condenser

Table 11-4 Main Technical Data of the King's Lynn Combined-Cycle Power Plant

Plant configuration:		GUD 1S.94.3
Number of blocks		1
Number of Gas Turbines / block		1
Main fuel		Natural gas
Ambient conditions:		
Ambient temperature	°C / °F	7.5/ 46
Ambient pressure	mbar	1,013
Cooling Type		Air cooled condenser
Performance data per block:		
Total Fuel Input to Gas turbine	MW	616
Gross Output	MW	347
Gross efficiency (LHV)	%	56.3
Gross Heat rate (LHV)	kJ/kWh	6,394
Heat rate (LHV)	Btu/kWh	6,060
Process heat	MJ/s	0

Due to the single-shaft concept, the arrangement of the main structures is longitudinal. The long turbine building is located between the enclosed HRSG at one end and the air-cooled condensers at the other. The building houses both the gas and steam turbines with the generator placed in between them. The steam turbine can be decoupled from the generator using a synchronous clutch (Figure 11-9).

The Taranaki Combined-Cycle Plant, New Zealand

The 360-MW Taranaki combined-cycle power plant is owned by Stratford Power, Ltd., New Zealand's first independent power producer, and supplies electricity as a baseload facility connected to the North Island transmission grid.

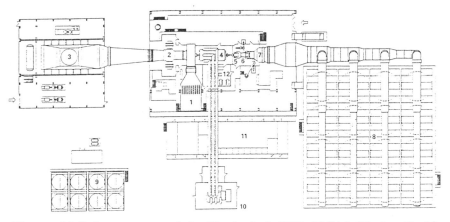

Figure 11-9 Arrangement of the Single-shaft GUD 1S.94.3 Block with Air-cooled Condenser (plan view)

The plant is located about 50 km from New Plymouth on a site 270 meters above sea level. Initially the plant will fire only natural gas taken from the Maui gas field but the design enables conversion to fire oil in the future should this prove necessary.

The power plant consists of a single-shaft block with one ABB GT26 gas turbine, one ABB natural circulation HRSG, and one ABB two-casing steam turbine, as well as the necessary auxiliary equipment. Due to the lack of cooling water on the site a wet cooling tower has been installed (Fig. 11-10).

Figure 11-10 View of Taranaki Combined-Cycle Plant

A high-efficiency triple-pressure reheat cycle was chosen for the plant, with HP live-steam at 103 bar (1,479 psig) / 568°C (1,054°F), reheat live steam at 24 bar (333 psig) / 568°C (1,054°F), and LP steam at 4.0 bar (44 psig) / saturated. The gas is preheated to further increase the efficiency (Fig. 11-11).

Table 11-5 Main Technical Data of the Taranaki Combined-Cycle Power Plant

Plant configuration:		KA26-1
Number of blocks		1
Number of Gas Turbines / block		1
Main fuel		Natural gas
Ambient conditions:		
Ambient temperature	°C / °F	11.6 / 53
Ambient pressure	mbar	981
Relative humidity	%	84
Cooling Type		Cooling tower
Performance data per block:		
Total Fuel Input to Gas turbine	MW	615
Gross Output	MW	359.9
Gross efficiency (LHV)	%	58.5
Auxiliary consumption and losses	MW	6.1
Net Power Output	MW	353.8
Net Efficiency (LHV)	%	57.5
Heat rate (LHV)	kJ/kWh	6,261
Heat rate (LHV)	Btu/kWh	5,934
Process heat	MJ/s	0

The Taranaki plant follows ABB's standard single shaft arrangement with the generator located in between the gas turbine and steam turbine (Fig. 11-12). The steam turbine can be decoupled from the gen-

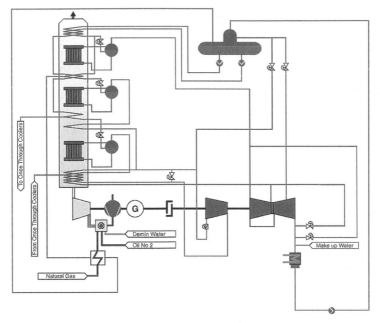

Figure 11-11 Process Diagram of Taranaki Combined-Cycle Power Plant

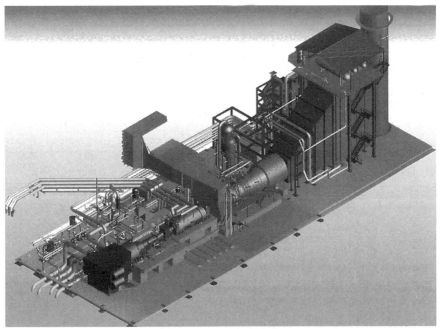

Figure 11-12 Arrangement of Taranaki Combined-Cycle Power Plant

erator by means of a self-shifting clutch for start-up and shutdown. The condenser is axial to the steam turbine—both are at ground level—eliminating the need for a steam turbine table. The hydrogen-cooled generator is mounted on a skid, together with its lube oil system and hydrogen coolers. The skid is mounted on transverse rails so it can be moved to the side for rotor inspections. The arrangement is very compact, with a foot print of the combined-cycle block measuring about 38 by 96 meters.

This plant will be the first in a series of ABB-standard reference single-shaft units for the 50 Hz market, for which there are already follow-up orders in the United Kingdom, Chile and Japan.

ABB will also be responsible for the operation and maintenance of the plant over the first six years. An online monitoring system will connect the plant to the ABB headquarters in Switzerland, where direct assistance will always be available to the plant operators.

The 121 MW IP-Cogen Combined-Cycle Cogeneration Plant, Hemeraj, Thailand

The IP-Cogen combined-cycle plant (Figure 11-13) produces both electrical power to the EGAT grid and process heat to an industrial complex. The plant provides 119 MW electrical output while generating 78 MJ/s of process steam. If no process steam is provided the power output will be 136 MW.

Dry low-NO_x burners reduce the exhaust emissions to a low 25 ppm NO_x (15% O_2 dry) when the plant fires its main fuel, natural gas. The process steam extraction reduces the heat dissipated to the environment significantly, keeping the impact to the environment low.

The IP-Cogen plant is made up of two ABB identical single-shaft blocks for the GT8C. Each block consists of one ABB GT8C gas turbine, one ABB CE natural-circulation HRSG, and one ABB single-casing extraction/condensing steam turbine. The gas turbine and the steam turbine drive a common generator installed between them. While the gas turbine is rigidly coupled to the generator, the steam turbine is equipped with a self-shifting and synchronizing clutch enabling start-up of the gas turbine independent from the steam turbine. The steam turbine clutch engages automatically during start-up as soon as the speed of the steam turbine reaches that of the generator.

Figure 11-13 View of IP-Cogen Combined-Cycle Cogeneration Power Plant

The cycle is a simple dual-pressure non-reheat cycle. The HP and LP steam feed the steam turbine which exhausts into an axially arranged condenser. The make-up water that replaces the losses associated with process steam export is supplied to the deaerator which is integrated into the condenser. The condensate is fed directly to the LP drum of the HRSG via the LP economizer. This means that no separate feed water tank is necessary making the cycle simple and easy to operate, (Figure 11-14).

A dual-pressure cycle is chosen for the IP-Cogen plant, in which HP steam of 63 bar (899 psig)/498°C (928°F) and LP steam of 5 bar (58 psig)/230°C (446°F) is generated in the HRSG before being fed to the steam turbine. The hot exhaust gases of the gas turbine, 531°C (988°F) are cooled to a temperature below 100°C (212°F) before leaving the HRSG through the stack.

Process steam is extracted from the steam turbine at 18 bar (247 psig) and supplied to the steam host. The entire plant is cooled by a

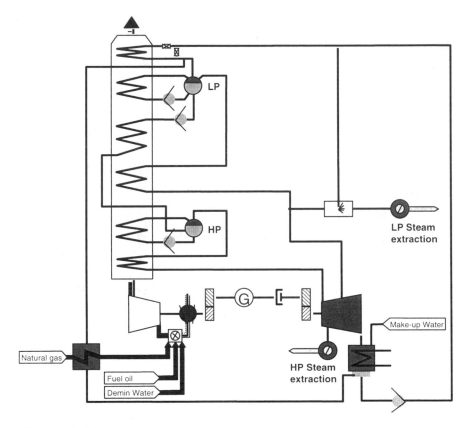

Figure 11-14 Flow Diagram of IP-Cogen Combined-Cycle Cogeneration Plant

cooling system employing wet cooling towers to dissipate the heat to the atmosphere. Table 11-6 summarizes the performance data of the IP-Cogen combined-cycle cogeneration plant. As shown, 69% of the fuel energy is converted into useful energy, with NO_x emissions below 25 ppm (15% O_2 dry), when firing natural gas.

Table 11-6 Main Technical Data of the IP-Cogen Combined-Cycle Cogeneration Plant

Plant configuration:		2 x KA8C-1	
Number of blocks		2	
Number of gas turbines / block		1	
Main fuel		Natural gas	
Ambient conditions:			
Ambient temperature	°C/ °F	28 / 82	
Ambient pressure	mbar	1,013	
Relative humidity	%	77	
Cooling type		Cooling tower	
Performance data per block:		**Cogeneration Mode**	**Power Generation Mode**
Total fuel input to the gas turbine	MW	142.1	142.1
Gas turbine power output	MW	46.9	46.9
Steam turbine power output	MW	13.8	22.7
Gross output	MW	60.7	69.6
Gross efficiency (LHV)	%	42.7	49.0
Auxiliary consumption and losses	MW	1.4	1.5
Net Power Output	MW	59.3	68.1
Net Efficiency (LHV)	%	41.7	47.9
Heat rate (LHV)	kJ/kWh	8,627	7,512
Heat rate (LHV)	Btu/kWh	8,177	7,120
Process heat	MJ/s	38.8	0
Fuel utilization (LHV)	%	69	—

The IP-Cogen plant is arranged outdoors. The gas turbine, steam turbine, and generator are in an enclosure suitable for the environment. The HRSGs are of an outdoor type. A compact arrangement provides short connections between the different plant areas (Fig. 11-15).

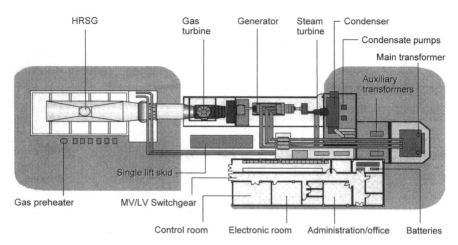

Figure 11-15 General Arrangement of IP-Cogen Combined-Cycle Cogeneration Plant

The 480 MW Combined-Cycle Power Plant, Monterrey, Mexico

The Monterrey combined-cycle power plant has been ordered by Comision Federal de Electricidad (CFE), the national utility of Mexico and will achieve commercial operation early in the year 2000. The plant, located in Monterrey (Nuevo Leon), is designed for pure electrical power generation with a net output of 484 MW and 56% net efficiency, making it one of the most efficient thermal power plants in Mexico.

The plant is based on ABB's KA24-1 standard reference plant and is among the first units of a series of installations in the 60 Hz market, applying the same design.

The gas turbines are equipped with the latest low NO_x technology to minimize emissions. Effluents are basically nil, qualifying it as a "zero discharge" site. An air-cooled condenser uses ambient air as heat sink to the cycle, keeping water consumption to a minimum.

The Monterrey power plant is made up of two identical 242 MW blocks, each comprising one 160 MW ABB GT24 gas turbine, one ABB CE once-through HRSG, and one ABB 90 MW double-casing reheat steam turbine unit. Modern combined-cycle plants employ a triple-pressure reheat cycle to achieve high efficiency, not so this plant.

The hot exhaust gases of the GT24 are used to generate steam at 160 bar (2,310 psig) for best efficiency. Producing steam at this pressure helps to avoid an IP system, therefore simplifying the cycle. The once-through HRSG avoids a thick walled HP Drum and results in a high thermal flexibility. The LP part is directly fed from the condenser (as seen previously in the IP-Cogen plant). Steam of 160 bar (2,310 psig) results in small steam volumes that can be efficiently expanded in the geared HP steam turbine. The barrel-type HP steam turbine design provides high thermal flexibility.

This simpler, dual-pressure, high-efficiency reheat cycle exceeds 56% net efficiency with an air-cooled condenser at 30°C (86°F) ambient temperature. The gas turbine and the steam turbine drive a common air-cooled generator installed between them. While the gas turbine is rigidly coupled to the generator, the steam turbine is equipped with a self-shifting and synchronizing clutch, enabling start-up of the gas turbine independent from the steam turbine. The steam turbine clutch engages automatically during start-up as soon as the speed of the steam turbine reaches that of the generator.

The cycle chosen for the Monterrey plant is a dual-pressure reheat cycle, wherein HP steam of 160 bar (2,310 psig)/565°C (1,049°F) and LP steam of 7 bar (87 psig)/320°C (608°F) is generated in the HRSG to be fed into the steam turbine. After partial expansion of the HP steam in the steam turbine, it is reheated in the HRSG at 37 bar, (522 psig) to 565°C (1,049°F).

The hot exhaust gases are thereby cooled from approximately 650°C (1,202°F) to below 100°C (212°F) before being exhausted through the stack. The entire power plant is cooled by an air-cooled condenser for the steam/water cycle and air blast coolers for the cooling system, to remove the heat from the plant auxiliaries (Fig. 11-16).

Table 11-7 summarizes the performance data of the Monterrey combined-cycle power plant. As shown in the table, more than 56% of the fuel energy is converted to electrical power, despite the air-cooled condenser operating at 30°C (86°F) ambient temperature that reduces the steam turbine output and increases auxiliary consumption. By this

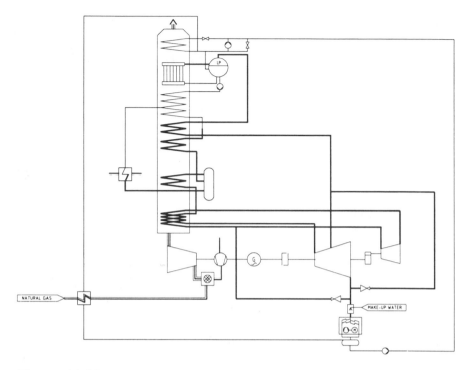

Figure 11-16 Process Diagram for Monterrey Combined-Cycle Power Plant

means, plant water consumption is kept to a minimum and no water body is heated up by the dissipated heat. NO_x levels in the exhaust are 25 ppm.

Each gas turbine, and the corresponding steam turbine, is arranged indoors in a common machine house, with sufficient lay-down area for inspections. The two blocks are fitted in separate buildings. The HRSGs are arranged outdoors. The balance of the plant is arranged to provide short pipe and cable routing to the different balance of plant equipment (Fig. 11-17).

Table 11-7 Main Technical Data of the Monterrey Combined-Cycle Power Plant

Plant configuration:		2 x KA24-1
Number of blocks		2
Number of Gas Turbines / block		1
Main fuel		Natural gas
Ambient conditions:		
Ambient temperature	°C /°F	30 / 86
Ambient pressure	mbar	969
Relative humidity	%	60
Cooling type		Air-cooled condenser
Performance data per block:		
Total fuel input to the gas turbines	MW	431.3
Gross output	MW	249.9
Gross efficiency (LHV)	%	57.9
Auxiliary consumption and losses	MW	7.8
Net Power Output	MW	242.1
Net Efficiency (LHV)	%	56.1
Heat rate (LHV)	kJ/kWh	6,413
Heat rate (LHV)	Btu/kWh	6,078
Process heat	MJ/s	0

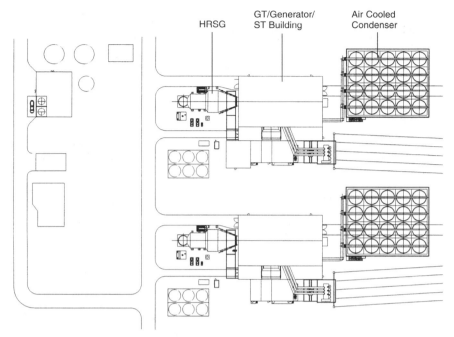

Figure 11-17 Layout of the Monterrey Plant Combined-Cycle Power Plant

CONCLUSION

CHAPTER **12**

12

Conclusion

The thermodynamic advantages of the combined-cycle concept enable efficiencies to be reached far above those of other types of thermal power plants. This technology above all others can fully exploit the high-temperature potential of modern gas turbines and the low-temperature "cold end" of the steam cycle. Coupled with low investment and operating costs, short delivery times and high operational flexibility, these factors ensure an overall low cost of electricity.

The thermodynamic advantages are also beneficial in cogeneration applications, especially where high electrical output is required because of the stable electrical output contribution of the gas turbine, which is not influenced by the steam process. The possibility of supplementary firing in the HRSG provides even greater operational flexibility where variations in the steam extraction demand are required.

Combined-cycle plants are suitable for daily cycling operation due to short start-up times and for continuous baseload operation. Part-load efficiencies are also high due to the control of the gas turbine inlet mass flow using inlet guide vanes.

Combined-cycles can be cooled by a cooling tower, a direct-cooling system or air-cooled condensers ensuring a wide range of applications. Where water is scarce, they are advantageous because the cooling requirement per unit of electricity produced is low due to the fact that the main cooling requirement applies only to the steam process (one third of total output).

Combined-cycle technology is one of the most environmentally acceptable large-scale power plant technologies in use today. Fuelled mainly by natural gas, they are ideally suited for use in heavily populated regions due to low emission levels for pollutants such as nitrous oxides and carbon monoxide/dioxide. High efficiency means less fuel must be burned for each unit of electricity produced, which also contributes to these low emission levels.

Development in combined-cycle plants continues through the advancement of the constituent components, which are now being developed specifically for use in combined-cycle applications. This has influenced the gas turbine in particular, with a trend towards pressure ratios and turbine inlet temperatures optimized for combined-cycle applications and a move towards more innovative machine concepts such as sequential combustion.

Although the fuel flexibility of combined-cycle plants is limited to gases and some oils this is becoming less significant as global gas distribution increases bringing natural gas to many countries which do not have their own natural reserves. The fuels that can be fired are those which are widely available in most parts of the world.

A wide range of combined-cycle power plant concepts is available and selection of a cycle concept is made according to the criteria of a specific project. Now that the combined-cycle is established as one of the main global power generation technologies, ideal cycle solutions are emerging matched to certain sets of criteria. Manufacturers have recognized in this a potential for the development of a range of standard plants. This leads to shorter delivery times, faster permitting possibilities, and lower risk to the investor due to proven components.

These and other factors point to a continuing dependence on combined-cycle technology for generating a main part of the world's electrical power well into the future.

APPENDICES

Conversions

Conversion of the main units used

Multiply	by	to obtain
bar	14.5	psia (psig=psia-14.5)
Btu	1.055	kJ
ft	0.30480	m
gal (USA)	3.7854	l
inch	2.54	cm
kJ	0.94781	Btu
kg	2.20046	lb
l	0.26417	US gal
m	3.2808	ft
psia	0.069	bar

Conversion formulae

To convert	Into	formula
°C	°F	(9/5)°C + 32
°F	°C	5/9 (°F -32)
K	°C	K - 273,15
K	°R	1.8 K

CALCULATION OF THE OPERATING PERFORMANCE OF COMBINED CYCLE INSTALLATIONS

When determining the performance data of a combined cycle the gas turbine is normally given and only the HRSG and steam turbine can be individually calculated. This appendix show the main steps involved in such a calculation process, starting with the HRSG.

1. Equations for the heat exchangers of the HRSG

The equations of energy, impulse and continuity are used to calculate the steady-state behavior of economizers.

The continuity equation comes down in the steady state to:

$$\Sigma m = 0 \tag{A-1}$$

The impulse equation can be simplified into:

$$\Delta p = f \, (\text{geometry}) \tag{A-2}$$

However, because the pressure loss both in the economizer and in the evaporator has a negligible influence on the energy equations, the assumption

$$\Delta p = 0 \tag{A-3}$$

is valid. In this case, the pressures along the heat exchanger remain constant, on both the gas and water sides. The energy equation for a small section dx of a heat exchanger, which can be treated approximately as a tube, can be written as follows:

$$dQ = k \cdot \Delta t \cdot \pi \cdot d \cdot dx \tag{A-4}$$

If it is assumed that the heat transfer coefficient k remains constant over the entire length of the heat exchanger (economizer or evaporator), Equation A-4 becomes:

$$\dot{Q} = k \cdot S \int_0^L \Delta t(x) \qquad \text{(A-5)}$$

In the general case, the expression cannot be integrated. The heat exchanger must therefore be dealt with in small elements.

In the special cases of a heat exchanger with counter or parallel flow, however, integration is possible assuming that the specific heat capacities of both media along the heat exchanger remain constant.

The result of the integration is the logarithmic average value for the difference in temperature, which can be written in the form:

$$\int_0^L \Delta t(x) = \frac{\Delta t_{Inlet} - \Delta t_{Outlet}}{In\left(\dfrac{\Delta t_{Inlet}}{\Delta t_{Outlet}}\right)} = \Delta t_m \qquad \text{(A-6)}$$

This average value can also be used for a superheater, an evaporator or a recuperator. The heat exchangers do not, in fact, operate in accordance with an ideal counterflow principle, but the errors remain negligible.

Substituting Equation A-6 into Equation A-5 yields:

$$\dot{Q} = k \cdot S \cdot \Delta t_m \qquad \text{(A-7)}$$

From Equation A-1, the amount of heat exchanged can be expressed as follows:

$$\dot{Q} = \dot{m}_S \cdot \Delta h_S = \dot{m}_G \cdot \Delta h_G \qquad \text{(A-8)}$$

At the design point, Equations A-7 and A-8 become:

$$\dot{Q}_0 = k_0 \cdot S \cdot \Delta t_{m0} \qquad \text{(A-9)}$$

$$\dot{Q}_0 = \dot{m}_{S0} \cdot \Delta h_{S0} = \dot{m}_{G0} \cdot \Delta h_{G0} \tag{A-10}$$

Dividing Equation A-7 by Equation A-9 and Equation A-8 by A-10 yields the formulae:

$$\frac{\dot{Q}}{\dot{Q}_0} = \frac{k \cdot \Delta t_m}{k_0 \cdot \Delta t_{m0}} \tag{A-11}$$

$$\frac{\dot{Q}}{\dot{Q}_0} = \frac{\dot{m}_G \cdot \Delta h_G}{\dot{m}_{G0} \cdot \Delta h_{G0}} \tag{A-12}$$

Subtracting Equation A-12 from A-11 produces:

$$\frac{\Delta t_m}{\Delta t_{m0}} = \frac{k_0 \cdot \dot{m}_G \cdot \Delta h_G}{k \cdot \dot{m}_{G0} \cdot \Delta h_{G0}} \tag{A-13}$$

This is the non-dimensional, global equation of heat transfer for the heat exchanger. If, in addition, Equation A-8 is taken into consideration and the heat transfer coefficient k is known, a system of equations is obtained that defines the heat exchanger.

2. Finding the heat transfer coefficient of HRSG sections

The heat transfer coefficient can be calculated using the following equation:

$$k = \frac{1}{\dfrac{1}{\alpha_G} + \dfrac{d_1}{2 \cdot \lambda} \cdot ln\left(\dfrac{d_1}{d_2}\right) + \dfrac{d_1}{d_2 \cdot \alpha_S}} \tag{A-14}$$

However, the relative values k/k_0 appear in the heat transfer equation. From this:

$$K = \frac{k}{k_0} = \frac{\dfrac{1}{\alpha_{G0}} + \dfrac{d_1}{d_2 \cdot \alpha_{S0}}}{\dfrac{1}{\alpha_G} + \dfrac{d_1}{d_2 \cdot \alpha_S}} \qquad \text{(A-15)}$$

The heat transfer coefficients on the gas end of the economizer and the evaporator (α_G) are from 0.1 to 0.01 times as large as those on the steam end (α_S). Moreover, both values always shift in the same direction (++, --)

For these reasons, the following relationship can be used:

$$K = \frac{k}{k_0} = \frac{\alpha_G}{\alpha_{G0}} \qquad \text{(A-16)}$$

The a-value on the gas end can be calculated as follows using the Nusselt number:

$$Nu_G = c \cdot Re^m \cdot Pr^n = \frac{\alpha_G \cdot d_1}{\lambda_G} \qquad \text{(A-17)}$$

Here, c, m, and n are constants that depend mainly upon the geometry involved. From this, the following expression is obtained:

$$\alpha_G = c' \cdot \lambda_G \cdot Re^m \cdot Pr^n \qquad \text{(A-18)}$$

If this is substituted into Equation (A-16), the geometric constant c' disappears:

$$K = \frac{\lambda_G \cdot Re^m \cdot Pr^n}{\lambda_{G0} \cdot Re_0^m \cdot Pr_0^n} \qquad \text{(A-19)}$$

For gases, the Prandtl number is almost exactly a constant. Therefore:

$$K = \frac{\lambda_G}{\lambda_{G0}} \cdot \left(\frac{Re}{Re_0}\right)^m \qquad \text{(A-20)}$$

For the Reynolds number, the following expression applies:

$$Re = \frac{c_G \cdot \rho_G \cdot d_1}{\mu_G} \qquad \text{(A-21)}$$

By substituting $\dot{m}_G/S$ for $c_G\,\rho_G$, one obtains:

$$Re = \frac{\dot{m}_G \cdot d_1}{\mu_G \cdot S} \qquad (A\text{-}22)$$

Then, substituting this expression into Equation (A-20), the geometric parameters disappear:

$$K = \frac{\lambda_G}{\lambda_{G0}} \cdot \left(\frac{\dot{m}_G \cdot \mu_{G0}}{\dot{m}_{G0} \cdot \mu_G}\right)^m \qquad (A\text{-}23)$$

If the mass flow is constant, all that remains is:

$$K = \frac{\lambda_G}{\lambda_{G0}} \cdot \left(\frac{\mu_{G0}}{\mu_G}\right)^m \qquad (A\text{-}24)$$

For m, one can use 0.57 for pipes that are offset from and 0.62 for pipes that are lined up with one another.

The value of the expression $\dfrac{\lambda_G}{\lambda_{G0}} \cdot \left(\dfrac{\mu_{G0}}{\mu_G}\right)^m$,

does not vary greatly and depends mainly on the properties of the gas. It can be replaced with the following approximation:

$$\frac{\lambda_G}{\lambda_{G0}} \cdot \left(\frac{\mu_G}{\mu_G}\right)^m = 1 - (\bar{t}_0 - \bar{t}) \cdot 5 \cdot 10^{-4} \qquad \text{(in SI-Units)} \qquad (A\text{-}25)$$

$\bar{t}_0$ and are $\bar{t}$ the average gas temperature along the heat exchanger in the design and operating point. This produces the relative value of K:

$$K = \left(\frac{\dot{m}_G}{\dot{m}_{G0}}\right)^m \cdot [1 - (\bar{t}_0 - \bar{t})] \cdot 5 \cdot 10^{-4} \qquad (A\text{-}26)$$

In this equation, only the exponent m depends to a slight extent on the geometry of the boiler.

It is more complicated to calculate an exact value for K in the case of a superheater because the heat transfer on the steam end is poorer than that in the evaporator.

When all of these equations have been obtained for all parts of the boiler, the HRSG has been defined mathematically. Similar equations can also be formulated for calculating the condenser.

3. The Steam Turbine

Most steam turbines in combined-cycle plants operate in sliding pressure mode and have no control stage with nozzle groups. This simplifies calculations, because simulation of the control stage and the inlet valves is fairly complicated.

A portion of a steam turbine with no extraction is defined by one equation for its swallowing capacity and one for its efficiency. The swallowing capacity can be defined using the Law of Cones (ellipse law).

$$\frac{\dot{m}_S}{\dot{m}_{S0}} = \frac{\bar{v} \cdot \rho_\alpha}{\bar{v}_0 \cdot \rho_{\alpha 0}} \sqrt{\frac{p_{\alpha 0} \cdot v_{\alpha 0}}{p_\alpha \cdot v_\alpha}} \cdot \sqrt{\frac{1 - \left(\dfrac{p_\omega}{p_x}\right)^{\frac{n+1}{n}}}{1 - \left(\dfrac{p_{\omega 0}}{p_{\alpha 0}}\right)^{\frac{n+1}{n}}}} \qquad \text{(A-27)}$$

In condensing steam turbines, the pressure ratio is always very small due to the low pressure at the steam turbine exhaust p_ω, . If simplified,this makes it possible to replace the quadratic expression with 1. The ratio of the absorption capacities is likewise close to 1.

What remains is then:

$$\frac{\dot{m}_S}{\dot{m}_{S0}} = \sqrt{\frac{p_\alpha \cdot v_{\alpha 0}}{p_{\alpha 0} \cdot v_\alpha}} \qquad \text{(A-28)}$$

At a constant rotational speed, the efficiency of a stage depends only upon the enthalpy drop involved. In part-load operation, however, no relatively great change occurs in that gradient except in the last stages. Because this means that the greatest portion of the machine is operating at a constant efficiency, it can be assumed that the polytropic efficiency remains constant. The turbine efficiency is calculated in the same way as for the design point.

The following formulae are used to calculate efficiency:

For parts of the turbine operating in the superheated zone:

$$\eta_{pol,dry} = \text{constant}$$

For parts in the wet steam region:

$$\eta_{pol} = \eta_{pol,dry} - \frac{(1 - x_\alpha) + (1 - x_\alpha)}{2} \qquad \text{(A-29)}$$

The polytropic efficiency selected should be such that the design power output is once again actually attained in the design point.

The following equation is used to determine the isentropic efficiency:

$$\eta_{is} = \frac{1 - \left(\dfrac{p_\omega}{p_\alpha}\right)^{\frac{x-1}{x} \eta_{pol}}}{1 - \left(\dfrac{p_\omega}{p_\alpha}\right)^{\frac{x-1}{x}}} \qquad \text{(A-30)}$$

These equations make it possible to establish the expansion line of the steam turbine. The power output of the steam turbine can be determined from this by allowing for dummy piston, steam turbine exhaust losses, mechanical and generator losses. The actual losses are given bases on the steam turbine type, size and live steam pressure. As a guidance the following losses can be considered:

- Piston losses
  For reaction type of steam turbines the piston losses account for 400 to 1000 kW mechanical equivalent losses. On the other hand impulse type of steam turbines, normally geared steam turbine, would only have gland steam losses accounting for approx. 0.2% mechanical equivalent losses

- Steam turbine exhaust losses
  Steam velocity at the steam turbine exhaust cause enthalpy losses, generally in the range of 20 to 35 kJ/kg

- Mechanical losses
  Bearing losses and other mechanical losses normally account for 0.3% mechanical losses

- Generator losses

The generator efficiency varies from 98 to 99% dependent on the steam turbine size

4. Solving the System of Equations

Taken together, all the equations in the HRSG, the steam turbine, etc. produce a system which can only be solved by iteration.

The following values are known:

- thermodynamic data at the design point
- the marginal conditions for the particular operation to be calculated (exhaust data for the gas turbine, cooling water data, etc.)
- operating mode of the feedwater tank (sliding or fixed pressure)
- Gas and Steam Tables

The following information must be found

- behavior of the steam cycle

Fig. A-1 shows the method used to find the solution. Starting with the superheater a first estimate for live steam temperature and pressure is made. Using the Law of Cones and the energy equation, the live steam flow and the gas temperature after the superheater are found. Next, from the heat transfer equation, a new value for live steam temperature can be determined. This is then used for further iteration. The procedure is repeated until all three equations have been solved.

The energy and heat transfer equations for the economizer and the evaporator can be used to determine a second approximation for live steam pressure.

If the feedwater tank is in sliding pressure operation, a first estimate for feedwater temperature is also necessary.

The new value obtained for live steam pressure is then used to continue calculation of the superheater and the turbine until all equations for the boiler and the Law of Cones agree. The next step is to calculate the preheating of the feedwater. This is used - if the pressure in the feedwater tank varies - to find a new approximation for feedwater

1. APPROXIMATION

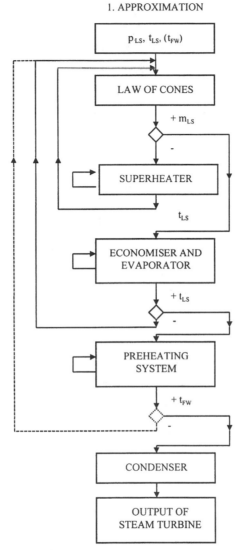

Figure A-1 Calculation of Operating and Part-Load Behavior: Method for Solving the System of Equations

temperature. The boiler is then recalculated, using this new value. Finally, the condenser pressure and extraction flow are determined in another iteration. Then, from this information, one can determine the power output of the steam turbine can be determined.

Symbols Used

c	Velocity
d	Diameter of a tube or pipe
Δh	Enthalpy difference
k	Heat transfer coefficient
$\dot{m}$	Mass flow
Nu	Nusselt number
p	Pressure
Δp	Difference in pressure
Pr	Prandtl number
$\dot{Q}$	Heat flow, amount of heat
Re	Reynolds number
S	Surface area
t	Temperature in °C
Δt	Difference in temperature
υ	Specific volume
x	Steam content of the wet steam
α	Heat transfer coefficient
η	Efficiency
μ	Dynamic viscosity
λ	heat conductance
$\bar{v}$	Average swallowing capacity of the turbine
ρ	Density

Indices Used

A	Air
FW	Feedwater
G	Flue gas
IS	Isentropic
LS	Live steam
Pol	Polytropic
S	Steam
0	Design point
1	Inlet / inside
2	Outlet / outside
α	Section of a turbine: inlet
ω	Section of a turbine: outlet

Bibliography

Bachmann, R., Fetescu, M., Nielsen, H.: "More than 60% Efficienty by Combining Advanced Gas Turbines and Conventional Power Plants, PowerGen 95, Americas.

Baerfuss, P.A., Vonau, K.H.: "GT8C single shaft combined-cycle: First Application of Industrial Power company Cogeneration at Eastern Industrial Estate, Thailand," Power-gen '96 ASIA, New Dehli, India (September 1996).

Bitterlich, E.: "Einflüsse der Thermodynamik kombinierter Dampf-Gas Prozesse auf die Konstruktion der Dampferzeuger," VDI 1971, No. 11, pp. 693 - 697 and No. 13, pp. 814 – 818.

Blad, J.: "Operation of a Combined District Heating and Power Production Scheme with a Combined-cycle Gas Turbine," von Karman Institute Lecture Series, 78/6.

Braun: "Dampferzeuger," Handbuchreihe "Energie", Vol. 5, Chapter 3.6 Performance.

Brückner, H., Wittchow, E.: "Kombinierte Dampf-Gasturbinen Prozesse: Einfluss aus Auslegung und Betrieb der Dampferzeuger," Energie und Technik 24 (1972), No. 5, pp. 147 - 152.

Brückner, H., Wittchow, E.: "Kombinierte Gas-/Dampfturbinen-prozesse: Wirtschaftliche Stromerzeugung aus Gas und Kohle," Brennstoff Wärme Kraft, Vol. 31 (1979), No. 5, pp. 214 –218.

Buchet, E.: "Les centrales a cycle combiné gaz-vapeur, Revue Generale de thermique," No. 233, May, 1981, pp. 391 - 404.

Buxman, J.: "Combined-cycles for Power Generation," von Karman Institute Lecture Series, 1978-6.

Charlier, J., Brückner, H.: "Combined Gas/Steam Turbine Power Plants with Supplementary Firing," Kraftwerk Union.

Cooper, V., Duncan, R.: "Baseload Reliability in a Combustion Turbine?" EPRI Journal, June, 1978.

Dibelius, G., Pitt, R. Ziemann, N.: "Die Gasturbine im Wandel technologischer u. wirtschaftlicher Entwicklungen," VGB Kraftwerkstechnik, 61 (1981), No. 2, pp. 75 - 82.

Endres, Dr. W., Gasturbine, Handbuchreihe "Energie", Vol. 3, Chapter 6.

Foster-Pegg, R.W., P.E., "The Steam-Turbocharged Gas Turbine," (unpublished paper, March 27, 1987). Address: 806 Columbia Ave., Cape May, N.J.

Foster-Pegg, R.W.: "Steam Bottoming Plants for Combined-cycles," Journal of Engineering for Power, April, 1978.

Frei, D.: "Gasturbinen zur Prozessverbesserung industrieller Wärmekraftwerke," Separate print, TR, Sulzer 4/75.

Frutschi, H.U., Plancherel, A.: "Comparison of Combined-cycles with Steam Injection and Evaporisation Cycles," 1988 ASME Cogen-Turbo IGTI, Vol. 3.

"GT26 kicks off New Zealand combined-cycle program." Reprint from Turbomachinery International, March-April 1997.

Gasparovic, N.: "Spitzenstromerzeugung in Gas/Dampfanlagen," Elektrizitätswirtschaft No. 24, Vol. 78 (1979).

Gericke, B.: "An- und Abfahrprobleme bei Abhitzesystemen insbesondere hinter Gasturbinen," Brennstoff Würme Kraft, Vol. 32, (1980 =, No. 11).

Gericke, B.: "Der natürliche Wasserumlauf in Ab-hitzedampferzeugern," Brennstoff Wärme Kraft, Vol. 39, 1979, No. 12.

Gratzki: "Messen und Regeln," Handbuchreihe "Energie", Vol. 4. Chapter 6.

Gusso, R., Moscogiuri, G.: "Combined Production of Thermal and Electric Power by Means of Gas Turbines and Combined-cycles," Quaderni Pignone 25.

Haase, Neidhöfer: "Elektrische Maschinen," Handbuchreihe "Energie", Vol. 4, Chapter 2.

Heins, G.L., Gurski, P.S.: "A low Btu coal gasification process for utility application," Modern Power Systems, April, 1981, pp. 37-40.

Hofer, P.: "Combined-cycle Yields Top Efficiency," Turbomachinery, March, 1981, pp. 30 - 42.

Hübner, R.: "Kombikraftwerke- umweltfreundliche Lösung des Energieproblems," Elektrische Energie-Technik, Vol. 23 (1978), No. 6.

Huhle, D.: "Combined-Cycle Plants for Generating Economical Medium-Load Power," Brown Boveri Review 61, 1974 (1), pp. 9 - 15.

"IGCC Combined-cycle Power Plants:" ABB Power Generation publication PGT 97 E.

Jaekel, G.: "Wärmeaustauscher in Gas-Turbinenkraftwerken," "Energie," Vol. 28, No. 1, January, 1976.

Jury, Dr. W.: "Single-Shaft Power trains (SSPTs): Flexible and Economic Power Generation for the Future," PowerGen Asia, Hong Kong (September 1994).

Kehlhofer, R., Thompson, B.R., Greil, C.: "Coal Gasification Combined-cycle Plants" A Clean Way from Coal to Electricity,"
VGB Congress, Strassbourg, France, 1986.

Kehlhofer, R.: "A Comparsion of Power Plants for Cogeneration of
Heat and Electricity," Brown Boveri Review 67, 1980 (8), pp.
504 - 511.

Kehlhofer, R.: "Calculation of Part-Load Operation of Combined
Gas/Steam Turbine Plants," Brown Boveri Review 65, 1978
(10), p. 679.

Kehlhofer, R.: "Combined Gas/Steam Turbine Power Plants for the
Cogeneration of Heat and Electricity," Brown Boveri Review
65, 1978 (10), pp. 680 - 686.

Kiesow Dr. H.J., Mukherjee, D.: "The GT24/26 Family Gas Turbine:
Design for Manufacturing," British Engineer's Conference,U.K.
(July 1997).

Knizia, K.: "Die Thermodynamik des Dampfprozesses," 3rd. Ed.,
Vol. I, Springer Verlag.

Koch, H., Brühweiler, E., Strittmatter, W., Sponholz, H.J.: "The Development of a Dry Low Nox Combustion Chamber and the Results Achieved," International Congress on Combustion Engines, CIMAC, Oslo, June 3 - 7, paper T1.

Krieb, K.H., Ratzeburg, W.: "Übersicht über neue Kraftwerkstech-
nologien," Technische Mitteil., No. 3, Vol. 71, March, 1978.

Lampert, D.: "Low-Temperature Corrosion in Feed-Heaters Heated
by Flue Gas," Brown Boveri Review 65, 1978(10), pp. 691 -
695.

Lampert, D.: "The Effect of the Structure of Combined Gas/Steam
Turbine Plants Upon Their Reliability," Brown Boveri Review
65, 1979 (2), pp. 133 ffg.

Lang, R.P., Chase, D.L.: "Steam and Gas Turbine Combined-cycle Equipment Currently Available for Natural Gas Pipelines," ASME Publication 79-GM-114.

Leikert, K.: "Feuerungen für Dampferzeuger hinter Gasturbinen," Separate Print, Techn. Mitteilungen, No. 9/10, Vulkan Verlag, Essen, 1976, pp. 103 - 108.

Listmann, R., Searles, D.: "Efficiency, Flexibility and Economical Viability: The CC Generation in a Deregulated Market." Latin American Power, Caracas, Venezuela (May 1997).

Löffel, H., Schauff, P.: "Wirtschaftlichkeit eines Gasturbinenkombiblocks mit Abhitzekessel zur Energieversorgung eines Chemiewerkes," Brennstoff Wärme Kraft, Vol. 31 (1979), No. 3.

Lumut Combined-cycle Power Plant.: published by Segari Energy Ventures Sdn Bhd, Kuala Lumpur and IMTE, Switzerland, (1997).

Mayer, M.: "GUD: An Unfired Combined-cycle Approach to Energy Utilization," ASME Publication 79-GT-131, 1979.

Meyer-Kahrweg, H.: "Betriebserfahrungen mit den aufgeladenen Dampferzeugern von 340 t/h Dampfleistung," Vortrag, VGB Conference "Dampfkessel und Dampfkesselbetrieb," 1975.

Mills, Dr. R.G.: "Development and Testing of a Modular Gas Turbine/Steam Combined-cycle," CIMAC, 14th International Congress on Combustion Engines.

Mori, Y., Iwata, K., Kimura, H.: "Design of Modern Combined-cycle Power Plant," Technical Review, Oct. 1980.

Morris & Wilson, Leyland, Watson & Noble: "Combined-cycle Power Generation," New Zealand Energy Research and Development Committee, Report No. 14, Oct. , 1976.

Müller. R.: "Kohleveredlung zu Gas und Flüssigtreibstoffen," Siemens- Energietechnik 2 (1980), No. 7, pp. 227 - 235.

Oest, H.: "Comparison between the Combined-cycle and the HAT Cycle," thesis, Department of Heat and Power Engineering, Lund Institute of Technology, (August 1993).

Pfenninger, Dr.H.: "Das kombinierte Dampf-/Gasturbinen-Kraftwerk zur Erzeugung elektrischer Energie," Brown Boveri Mitt. 60 1974 (9), pp. .389 - 397.

Pickhardt, K.F.: "Abhitzekessel hinter Gasturbinen," "Energie," Vol. 30, No. 9, Sept., 1978.

Roberts, R., Balling, L.,Wolt, E.,Fränkle, M.: "The King's Lynn Power Station: The introduction of the advanced single shaft concept in the IPP market." Power-Gen Europe, Madrid, (June 1997).

Rohrer, A.: "Comparison of Combined Heat and Power Generation." ASME Cogen Turbo Expo, Vienna,(August 1995).

Schneider, A., Unseld, Dr. H.: "Ein kombiniertes Gasturbinen-Dampfkraftwerk für industrielle Versorgung," Energie, Vol. 27, No. 3, March 1975.

Schüller, K.H.: Heizkraftwerke, Handbuchreihe "Energie", Vol. 7, Chapter 4. Schüller, K.H.: Industriekraftwerke, Handbuchreihe "Energie" Vol. 7, Chapter 5.

Schwarzenbach, H., Koch, E.: "Dampfturbinen," Handbuchreihe "Energie", Vol. 3, Chapter 7.
Seippl, C, Bereuter, R.: "Zur Technik kombinierter Dampf und Gasturbinenanlagen," Brown Boveri Mitt. 47 1960 (12) pp, 788 – 799.

"The 347MW King's Lynn Single-Shaft Combined-cycle (GUD) Power Station with Air-Cooled Condenser," Siemens Power Generation publication No. A96001-U10-X-7600.

Thermodynamiques Combines Gaz/Vapeur Aspects Theoriques Applications Pratiques et Aspects d'Exploitation," AIM Liege, Centrales Electriques modernes - 1978.

Timmermanns, A.P.J.: "Combined-cycles and Their Possibilities," von Karman Institute Lecture Series 1978-6.

Tomlinson, L.O., Snyder, R.W.: "Optimization of STAG Combined-cycle Plants," presentation at the American Power Conference, Chicago III., Apr. 29 – May 1, 1974.

Traupel, W.: "Thermische Turbomaschinen," Springer Verlag.

Wadman, B.: "New High Efficiency Combined-Cycle-System," Diesel/Gas Turbine World-wide, Jul./Aug., 1980.

Warner, J., Nielsen H.: "A selection method for optimum combined-cycle design," ABB Review, 8/93.

Watzel, G.V.P., Essen: "Beeinflussung des Leistungsverhältnisses zwischen Gas und ampfturbine bei kombinierten Prozessen," Brennstoff Wärme Kraft, Vol. 22 (1970), No. 12.

Wunsch, A, Mayrhofer, M.: "Power Plants for the Medium Output Range, Criteria Governing the Choice of the Optimum Plant," Brown Boveri Review 65, 1978 (10), pp. 656 - 663.

Index

C

Calcium carbonate, 139
Capital costs, 11-12
Carbon dioxide emissions, 221,
227-228, 266
Carbon monoxide, 266
Carnot efficiency, 36-37, 107
Casing for HRSG, 170
Catalyst, 225-227
Categories (gas turbine),
157-160
Ceramics, 233
Closed-air circuit cooling,
181-182
Closed-loop control system,
184, 190-198
Closed-steam cooling, 234
Coal and gas cycle, 145-148
Coal-fired plant, 10, 16, 21, 28,
145-148, 227-228
Cogeneration, 7, 126-141,
246-250, 256-260, 266
  design parameters, 134-135
  district heating power plants,
  135-138
  evaluation of cycle, 129-134
  industrial power stations,
  127-129
  seawater desalination units,
    138-141
Cold-casing design, 170
Combined-cycle concepts, 2-3,
14-15, 30, 48-124
  basic concepts, 49-103
  selection of concept, 103-124
Combustion air conditions,
222-223

Combustion chamber, 43,
233-235, 240
Combustion pressure, 222
Combustion products, 220-229
Combustion with excess air, 238
Communication capability, 184
Competitive risks, 7
Competitive standing, 14-26
  availability/reliability, 23-25
  construction time, 25-26
  efficiency, 16-21
  fuel costs, 16-21
  maintenance costs, 22-23
  operation costs, 22-23
  turnkey prices, 15-16
Components, 156-188
  bypass stack, 187-188
  control system, 184-186
  cooling system, 186-187
  electrical equipment, 182-183
  gas turbine, 156-164
  generators, 180-182
  HRSG, 164-175
  steam turbine, 176-181
Compressor, 156-157, 162-163,
235
  fouling, 162-163
  size, 235
Concept selection, 103-124-
  defining requirements, 104
  site-related factors, 104-119
  solution determination,
    119-124
Condenser vacuum, 49
Condensing mode, 247
Construction cost, 11
Construction time, 25-26

Other titles offered by PennWell...

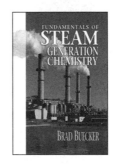
To purchase a PennWell book...

- Visit our online store www.pennwell-store.com, or
- Call 1.800.752.9764 (US) or +1.918.831.9421 (Intl), or
- Fax 1.877.218.1348 (US) or +1.918.831.9555 (Intl)

Branch out into other areas with our NONTECHNICAL BOOK SERIES!

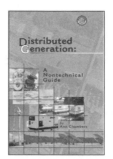
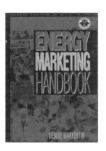

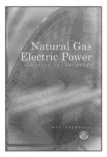
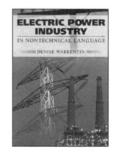
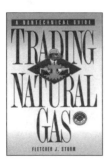